Managing Corporate Energy Needs: The Role of Management Accounting

by

Harold E. Wyman
Professor of Accounting
University of Connecticut

and

J. Edward Ketz
Assistant Professor of Accounting
Pennsylvania State University

A study carried out on behalf of the
National Association of Accountants
New York, New York

Published by

National Association of Accountants
919 Third Ave., New York, N.Y. 10022

Kathy Williams, Editor
Mandel & Wagreich, Inc., Cover

NAA Publication Number 82136

ISBN 0-86641-084-8

Foreword

The attention of management and the general public was directed toward energy issues in a sudden and dramatic fashion as a result of the oil embargo by the Organization of Petroleum Exporting Countries in 1973–74. Worldwide dependence on imported oil enabled OPEC to raise the price of crude oil beyond the point that anyone would have imagined—from $1.90 per barrel in 1972 to $32.00 per barrel in 1981.

U.S. industry faced a series of problems as a result of the energy situation. The increased cost of energy supplies had a direct impact on the purchaser, but the increased cost also spread through the economic system and was felt indirectly by many firms as an increase in the cost of their raw materials. Industry responded to the government's call to increase energy efficiency and made significant progress. Economic growth, however, is such that, even with the increased efficiency, expected future energy use will continue to grow.

The National Association of Accountants initiated this study in an effort to help corporations manage their energy needs more effectively and to indicate the role of the management accountant in this area. Included in this study you will find information on how to improve existing energy programs and how to start one if you have no program. The study includes information from a mail survey of 1,000 firms and from personal interviews with executives of 12 companies chosen to illustrate the approach to a variety of energy problems.

The study shows that there is considerable room for improving the management of corporate energy, based on the mail survey results. It emphasizes the need to apply sound management to energy needs in terms of strategic planning, capital expenditure analysis, and management control. Specific examples of how to do this are provided throughout.

Guidance in the preparation of this study was generously provided by the Project Committee:

Charles L. Grant (Chairman)
Becton, Dickinson & Co.
Rochelle Park, New Jersey

Milton F. Usry
Oklahoma State University
Stillwater, Oklahoma

Michael L. Ferrante
Ernst & Whinney
White Plains, New York

Theodore C. Gearhart
Arlington, Virginia

The report reflects the views of the researchers and not necessarily those of the Association, the Committee on Research, or the Project Committee.

Stephen Landekich
Research Director
National Association of Accountants

Acknowledgments

The completion of a research project such as this one involves the cooperation of many people whom we would like to thank. In particular, we are grateful for the outstanding assistance received from NAA Research Director Stephen Landekich. He made many valuable suggestions that we feel improved our work.

Because of our promise to keep the responses confidential, we cannot individually acknowledge all of the executives who took time from a busy schedule to answer our mail survey, but we would like to acknowledge that we could not have done it without them. We particularly would like to thank the following people for their help:

Spruill Bunn, Hardee's Food Systems, Inc.
David W. Clark, Lydall, Inc.
Barbara Dreyer, Providence and Worcester Railroad Co.
Richard S. Eckstrom, Southern New England Telephone Co.
Chet L. Helt, Pic'N Pay, Stores, Inc.
J. Russell Hill, Hardee's Food Systems, Inc.
Rex O. Lewis, Armstrong Rubber Co.
Edward F. Lyons, Champion International Corp.
A.H. Manchester, Hercules Inc.
Paul Marchi, Southern New England Telephone Co.
Ralph Marrone, Northeast Utilities
Howard Oldenberg, J.C. Penney Co., Inc.
Harold O'Tuel, Burlington Industries
James P. Riley, Allied Supermarkets
Richard Schenkel, American Motors Corp.
Carl F. Stenman, Connecticut General Life Insurance Co.
Bernie Thibado, Lydall, Inc.

Finally, we would like to thank Mrs. Connie Seymour of the University of Connecticut for cheerfully handling an infinite number of details which always had a deadline.

Harold E. Wyman
J. Edward Ketz

Table of Contents

List of Tables

Table		Page

Managing Corporate Energy Needs: The Role of Management Accounting

Executive Summary

The research in this study focuses on three questions: Do energy needs warrant particular attention by corporate management? Is there a role for management accounting in the energy management process? If so, what should that role include?

In March 1980, the National Association of Accountants issued a research prospectus calling for projects directed to a "thorough analysis of the impact of the energy problem on management accounting information related to monitoring and reducing energy consumption as well as budgeting for alternative energy sources." Topics to be addressed included:

1. Organization of the energy effort and establishment of an energy policy.
2. Development of internal and external perspectives.
3. Determination and evaluation of energy impact assumptions.
4. Assignment of priorities and the establishment and budgeting of an energy program.
5. Implementation and monitoring of an energy program.

The study presented here addresses these issues in the context of *managing* an energy program.

The first question, "Do energy needs warrant particular attention by corporate management?", is answered "yes" based on an analysis of the past, present and future energy situation presented in Chapter 1. The affirmative answer to the question, "Is there a role for management accounting in the energy management process?", is based on an analysis of the energy management process as it is presented by persons who have written on the subject (discussed in Chapter 2), and as it really exists, according to responses from a mail survey presented in Chapter 3.

The energy management process of strategic planning, capital expenditures, energy audits, and management control—planning, organizing and measuring—is analyzed in Chapter 2. From this analysis came questions that were included in a mail survey which established how extensively various management practices are used in a representative group of companies.

The role of management accounting in the energy management process was defined in two ways: *analytically,* by looking at the energy management process in Chapter 2, and *practically,* by identifying areas where management accounting should take a more active role, based on responses to the mail survey. Results of the mail survey, presented in Chapter 3, represent the responses of 255 firms from a stratified sample of 1,000 firms.

Specific energy management programs for 12 companies are summarized as case studies in Chapter 4. These cases, based on field visits, help verify the results of the mail survey and expand the survey of energy management programs beyond the questionnaire.

We found that management accountants need to take a more active role in the energy management process at their companies. Some company managements view energy as a technical problem rather than as a management problem, however, so they leave energy management to the technicians and engineers. The facts and figures that are used to show energy trends, recommend cost and energy savings, adjust future cash flows and the like, however, would be better left to the management accountants who massage such data daily. Some statistics from our analysis of firms responding to the mail survey regarding the role of management accounting in the energy management process—particularly as it relates to capital expenditures, energy audits and measuring energy costs—support our belief that management accountants should be more active in this area. For example:

- 48.2% identify the quantity of additional energy or energy savings connected with projects
- 56.1% adjust future cash flows connected with capital expenditure projects for anticipated increases in energy costs
- 42.5% conduct energy audits
- 48.4% have energy consumption reporting for management
- 37.2% prepare energy usage trend reports
- 57.1% have guidelines for measuring current energy consumption
- 34.5% separate the rate variances from the usage variances in analyzing performance against standard or budget

How to Use This Study to Improve an Energy Program

If a company already has an energy program, it can use this study to find new ideas. Ideas for energy strategies are presented in Table 6 (page 16). A survey of what other firms have done appears in Table 8 (page 25).

Ideas for improving the capital expenditure process can be found in Table 10 (page 28). Ideas for capital expenditure projects frequently come from energy audits, but only 42.5% of the firms in the survey conducted these. Those firms that want to improve their energy programs should engage in energy audits, which are discussed on pages 30–32.

Ideas for improving the energy management process can be obtained by analyzing the questions in Table 14 (page 33), Table 16 (page 35), Table 18 (page 38), and Table 20 (page 41). One can make a quick overall analysis of a company's energy program by comparing the existing program to the ten questions rated highest in the survey (page 94).

How to Start an Energy Program

If a company has done nothing in the area of energy management but wants to start as a result of reading this study, it should obtain the *Energy Conservation Program Guide for Industry and Commerce,* NBS Handbook 115, available from the Superintendent of Documents, U.S. Government Printing Office, Washington, D.C. 20402 (order by SD Catalog No. C13.11:115). This guide contains chapters on energy conservation program implementation, energy conservation opportunities, engineering data and factors, financial evaluation procedures, sources of assistance, safety and pollution considerations, employee participation, and energy flow measurements, plus a comprehensive bibliography. It would be an invaluable tool for the business that cannot afford expensive consultants.

Chapter 1

Analysis of the Past, Present and Future Energy Situation

Management and the general public became aware of energy issues in a sudden and dramatic fashion as a result of the 1973–74 oil embargo by the Organization of Petroleum Exporting Countries (OPEC). Without realizing it, both industry and the public had become increasingly dependent upon imported oil. For example, between 1960 and 1974, imports as a percentage of consumption increased from 19% to 38%.[1]

Not only did the embargo emphasize U.S. dependence on foreign oil, but it also showed the ability of OPEC to raise the price of world oil. We have our experience since 1974 plus projections of the future to help us determine if there is a need for companies to plan and control energy needs. A review of that experience and of our government's reaction provides background.

The federal government placed statutory limits on the prices of some energy sources, including "old oil," which resulted in a disruption of the market mechanism for equating supply and demand. Energy shortages were the consequence. Although President Reagan deregulated oil prices early in 1981, some price controls still exist under the Natural Gas Policy Act of 1978. Because of the conditions prevailing under price controls, the aggregate domestic supply of energy is not as large as it could be, and the government's role has been criticized.

Forecasts Show Need for Imports

Even with the deregulation of oil prices and the phased deregulation of the price of newly discovered natural gas under the Natural Gas Policy Act of 1978, it is not clear whether our supply will keep pace with the increasing demand forecast for the future. The energy experts of SRI International predict that U.S. energy demand will continue to grow through the year 2000 as shown in Table 1.

The forecast shows that demand will almost double between 1970 and 2000 when it will approach 60 million barrels of oil equivalent per day. A barrel of oil equivalent is the quantity of energy contained in a 42-gallon barrel of crude oil, which is equal to 5.8 million BTUs. The BTU, British Thermal Unit, is the quantity of heat required to raise the temperature of one pound of water by one degree Fahrenheit. Note that the earlier forecast of energy made in 1973 showed a much higher future demand than the 1978 forecast. The latter forecast was based on U.S. experience since 1973 and shows the modifying effect of energy conservation measures implemented as a reaction to the energy situation of 1973–74.

Not only is the total demand for energy expected to grow, but, according to one forecast, the amount of energy required to be imported will double between 1975 and the year 2000. This forecast, presented in Table 2, shows that even though significant increases are predicted

TABLE 1
Forecasts in 1973 and 1978 of U.S. Energy Demand in 2000

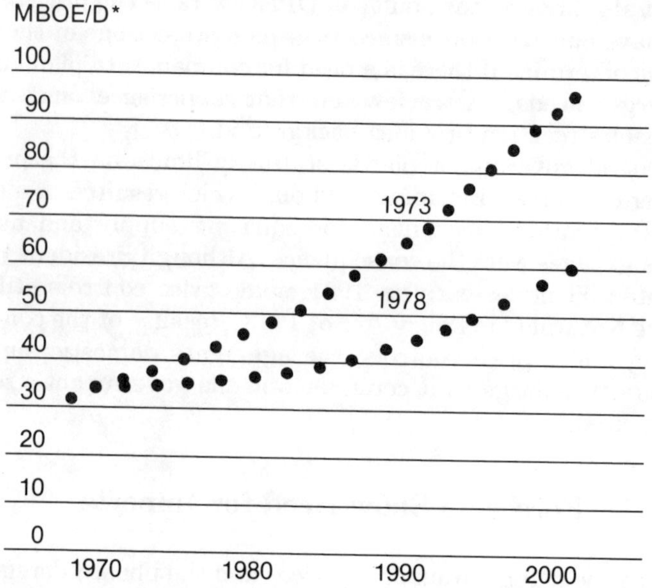

*Millions of barrels of oil equivalent per day.

Source: Henry, *et al.*[2]

TABLE 2
U.S. Energy Shortfall (quads)

	YEARS			
	1975	**1985**	**1990**	**2000**
Domestic Supplies:				
Natural gas and biomass	22.19	23.03	21.05	19.31
Petroleum	17.73	29.40	27.20	22.00
Coal and wastes	15.45	20.96	26.25	35.82
Hydro and geothermal	3.02	3.98	4.50	7.80
Nuclear	1.65	7.98	15.31	28.49
Other (solar and shale)	—	.32	.85	4.85
Total	60.04	85.67	95.16	118.27
Domestic Consumption	71.19	97.82	109.95	138.83
Shortfall to be satisfied by imports	11.15	12.15	14.79	20.56

Note: One quad equals 2.929×10^{11} kwh = 10^{15} BTU

Source: Belding[3]

in supplies of energy from coal and waste, hydro and geothermal, and nuclear, there still will be a steadily increasing shortfall.

It is not surprising that the United States will continue to need imported energy in the future as we have not demonstrated an ability to reduce our need for imported oil, a major energy source, despite the embargo of 1973–74 and our subsequent awareness of vulnerability.

Table 3 shows our increasing dependence on imported oil prior to 1974, but even after the disruptive effect of the oil embargo, both the quantity imported and its importance as a percentage of consumption continued to increase.

Escalation in Energy Prices

Another aspect of the oil embargo and the power of OPEC was the massive jump in the price of oil. Slight increases in the price from \$1.30 per barrel in 1970 to \$1.65 in 1971 to \$1.90 in 1972 were followed by a significant increase to \$2.70 in 1973 followed by an increase that more than tripled the price to \$9.76 in 1974. Since 1974, the price has tripled again to \$32 per barrel in June 1981. The record from 1970 through 1980 is shown in the list on page 8.

1970	$ 1.30	1976	$11.51
1971	1.65	1977	12.40
1972	1.90	1978	12.70
1973	2.70	1979	16.97
1974	9.76	1980	28.67
1975	10.72	June 1981	32.00

The effect of the escalation in energy costs stemming from increases in the price of crude have created serious problems for industry, according to a 1978 survey of the *Fortune* 1,000 largest industrial firms. In this survey of problems created by the energy crisis, respondents could rank their answer on a scale of 1—no problem to 5—serious problem. Over 50% marked the cost of energy supplies as a most serious problem. One can judge the severity of the various concerns about energy problems from a recap of the responses as shown in Table 4.

If we judge a rating of 4 or 5 on the response scale as indicating a serious problem, then the three most serious problems perceived by industry were: cost of energy supplies, 79%; uncertainty within the company caused by government actions in energy matters, 62%; and cost of raw materials, 59%.

Energy Efficiency Improving Since 1972

Despite problems caused by the federal government's lack of a clear

TABLE 3
U.S. Consumption of Oil

Year	Consumption	Production	Imports	Imports as a % of Consumption
1960	9.7 mb/d*	8.0 mb/d	1.8 mb/d	19%
1962	10.2	8.4	2.1	21
1964	10.8	8.8	2.3	21
1966	11.9	9.6	2.6	22
1968	13.0	10.6	2.8	22
1970	14.4	11.3	3.4	24
1972	16.0	11.2	4.7	29
1974	16.2	10.5	6.1	38
1976	17.0	9.7	7.3	43
1979**	18.4	10.0	8.4	46

*millions of barrels per day
**estimated

Source: Stobaugh and Yergin[4]

TABLE 4
Problems Created by the Energy Crisis

	Response in Percentages					
	Serious Problem			No Problem		
Problems	**5**	**4**	**3**	**2**	**1**	**No Opinion**
Cost of raw materials	33.2	25.4	25.4	6.6	7.0	2.4
Shortage of raw materials	13.0	9.6	22.6	29.3	21.3	4.2
Cost of energy supplies	50.2	28.7	13.0	6.1	1.2	0.8
Erratic delivery of raw materials	6.3	9.3	18.2	28.7	29.1	8.4
Shortage of energy supplies	18.0	15.9	27.8	16.3	20.4	1.6
Shortage of packing materials	1.3	4.7	9.8	19.7	50.4	14.1
Uncertainty within the company caused by government actions in energy matters	41.0	20.9	20.5	9.8	6.6	1.2
Maintaining customer service standards	3.8	4.2	14.6	21.4	46.0	10.0
Coordinating physical distribution and marketing (sales)	0.8	5.4	12.1	21.7	47.1	12.5
Worker layoffs, furloughs, or leaves of absence	1.3	5.9	17.6	21.7	48.9	4.6
Maintaining long-term supplier commitments on energy sources	8.4	19.2	22.6	19.7	24.3	5.8
Reduced working hours due to energy shortages	2.5	5.4	17.3	24.0	47.5	3.3

Source: Lambert and Stock[5]

energy policy, industry has responded to the government's challenge to increase energy efficiency, which is measured by relating energy consumed to production. The Energy Policy and Conservation Act, as amended by the National Energy Conservation Policy Act, set industrial efficiency targets for selected industries. These targets represented the percentage improvement that government experts felt could be achieved between 1972 and 1980 in the ten industries that consumed the most energy. These ten industries, their 1972 energy use, their anticipated 1980 use with no change in efficiency, 1980 improvement targets set by the government, 1980 energy use assuming efficiency is attained, and the projected 1980 savings are presented in Table 5.

If the government's targets were met, there would be a savings

TABLE 5
Potential Savings of Energy in 1980

SIC		1972 energy use (base year) 10^{12} BTU/yr.	1980 energy use with base year efficiency 10^{12} BTU/yr.	1980 improvement target percent	1980 energy use assuming attainment of net target 10^{12} BTU/yr.	Projected 1980 savings through attainment of net target 10^{12} BTU/yr.	Projected 1980 savings through attainment of net target BFOE per day*
20	Food and kindred products	1,047	1,195	12	1,052	143	62,200
22	Textile mill products	474	567	22	440	127	55,200
26	Paper and allied products	1,388	1,526	20	1,210	316	137,400
28	Chemical and allied products	3,087	4,800	14	4,128	672	292,200
29	Petroleum and coal products	2,993	4,007	12	3,527	480	208,700
32	Stone, clay, and glass products	1,462	1,753	16	1,478	275	119,600
33	Primary metal industries	4,246	5,167	9	4,690	477	207,400
34	Fabricated metal products	442	587	24	445	142	61,800
35	Machinery, except electrical	437	707	15	601	106	46,100
37	Transportation equipment	414	690	16	580	110	47,800
	Total	15,990	20,999		18,151	2,848	1,238,400

*Barrels fuel oil equivalent per day; conversion factor is 6.3×10^6 BTU

Source: Department of Energy[6]

equal to 1,238,400 barrels of fuel oil equivalent per day—an important saving. Notice, however, that industrial growth requiring energy is such that even with this savings from efficiency the expected energy use would still increase from 15,990 in 1972 to 18,151 in 1980, expressed in 10^{12} BTUs per year. Meeting the government's efficiency targets would not reduce energy consumption below levels of the past, but it would slow the growth in the demand for energy.

In the most recent report of the Department of Energy, all ten industries had made progress toward the targets, and six of the industries had attained their 1980 energy efficiency targets by December of 1978. Those attaining their targets were food, chemicals, petroleum, primary metals, machinery, and transportation equipment. Based on performance to the date of the report, the Department of Energy expected the other four industries to meet their target by 1980. As a result, the absolute level of energy consumption for the total of the ten industries had fallen 2.5% since 1972 during a period when output had risen 17%.

Long-Term Prospects

In the fall of 1981, as this was being written, there was a mood of optimism and talk of an oil "glut." Hundreds of oil tankers were idle. They were not needed because of our conservation efforts and world overproduction. The mail survey of *Fortune* firms presented earlier in Table 4 showed that a "shortage of energy supplies" was not a major concern; only about one-third of the respondents considered this a serious problem, ranking it 4 and 5 on the five-point scale. Yet, there is a hint that this situation is only temporary.

Certainly the information presented in Table 1 about future demand increasing, in Tables 2 and 3 about future imports increasing and the history of price increases from 1973 to 1981 do not support a feeling of long-term optimism about energy. In fact, we may not escape the decade of the '80s without another crisis. Franklin A. Lindsay, head of the Committee for Economic Development's Research and Policy Committee and vice chairman of the energy and raw materials subcommittee of the Industry Advisory Committee, Organization for Economic Cooperation and Development, in the fall of 1981, wrote:[7]

> "Recently, I conducted an informal survey of those knowledgeable in energy matters, Middle Eastern politics, and national security. Nearly everyone believed it will take at least ten years before the United States can significantly reduce its dependence on Middle Eastern oil.

No one thought there was less than a 50% chance of a cutoff during the 1980s. In fact, many said that the chance we will avoid another cutoff was less than 10%."

Given the preceding analyses of events since 1973 and forecasts, it follows that because of its overall effect, importance and outlook, energy will remain an important management consideration in the 1980s. Because energy is used for transportation, production, heating, air conditioning, lighting and communication, the energy situation affects a broad range of companies. The analysis in this chapter leads one to the conclusion that, yes, energy needs warrant particular attention by corporate management.

Notes

1. Robert Stobaugh and Daniel Yergin, "The Energy Outlook: Combining the Options," *Harvard Business Review* (January-February 1980), p. 58.
2. John P. Henry, Jr., V. Eugene Harless, and Jay B. Kopelman, "World Energy: A Manageable Dilemma," *Harvard Business Review* (May-June 1979), p. 153.
3. John A. Belding, "Alternatives to Oil and Gas through Energy Management," *Journal of Energy* (January-February 1978), p. 2.
4. Stobaugh and Yergin, *op. cit.,* p. 58.
5. Douglas M. Lambert and James R. Stock, "Organizing and Implementing the Corporate Energy Plan," *MSU Business Topics* (Summer 1979), pp. 5–15.
6. United States Department of Energy, *Industrial Energy Efficiency Program,* Annual Report July 1977 through December 1978 (December 1979), p. 2.
7. Franklin A. Lindsay, "Plan for the Next Energy Emergency," *Harvard Business Review* (September-October 1981), p. 153.

Chapter 2

The Energy Management Process and Questionnaire Design

The questions, "Is there a role for management accounting in the energy management process?" and "If so, what should that role include?", can be answered only by examining the energy management process. Because firms have been coping with the energy situation since 1973–74, one finds discussions of their experiences reported in various publications. A search of the relevant literature provided selected examples of how management planned, organized and controlled energy as well as suggestions for others. One cannot rely just on what has been reported, however, because it does not include a measure of the incidence or frequency of important components of the energy management process. This type of information must be gathered by a survey.

We supplemented the mail questionnaire with visits to 12 companies to see if there were aspects of energy management not covered in the questionnaire and to give a breadth to the survey by providing specific examples to support the findings of the survey. After reviewing the library material, we realized that there was a role for management accounting in the energy management process, so we included questions related to this role in the mail questionnaire.

Strategic Planning in the Energy Management Process

The energy management process may be organized into the following components:

- Strategic planning
- Capital expenditures
- Management control (goals, organization, measures)

The components represent general management concepts applied specifically to energy. For example, here is a list of topics that need to be considered in strategic planning for energy:

1. *Current energy consumption.* How can the firm best survey and measure its current energy consumption?
2. *Conservation goals.* Once the extent of energy usage is known, the firm will want to reevaluate whether it can be more efficient in this usage. Is it possible for the company to conserve energy? If it *is* possible to conserve energy, what realistic goals can be set that will not impair the firm's performance?
3. *Product line.* Each of the company's products needs to be reexamined in terms of its energy inputs. The firm needs to reassess the relative economies of the product, including the elasticity of its demand. What will be the cost of the product? Can those costs be contained?
4. *Production planning.* The corporation continually must probe for alternative ways to produce its goods. Can one redesign the manufacturing process with energy considerations in mind? Can the labor/capital mix be changed? Can a more energy-efficient production process be found?
5. *Transportation.* What does the corporation pay for the raw materials transported in and for transporting out the products sold? How can the firm reduce its transportation costs given that transportation involves high energy costs?
6. *Plant and office locations.* Companies will want to review the locations of their plants and offices, given the energy outlook. Product costs may be lowered; cheaper labor may be found to offset increased energy costs; more efficient transportation routes may be generated.
7. *Market planning.* The energy crisis has raised the energy consciousness of the consumers. Will they still want the same products, given the energy situation? Also, can the firm develop a market for services or products to help consumers reduce their energy needs?
8. *Financing.* As the energy crisis has led to greater energy costs, the total costs of the firm generally have risen. To what extent can the corporation finance these higher costs? Is equity or debt financing available, and how costly are the funds? Some firms use different costs of capital for energy projects than for other projects. Is it really desirable to use different costs of capital? If so, how should they differ?
9. *Contingency planning.* The future most likely will bring fuel shortages, raw materials shortages or other energy-related

disruptions. What contingency plans exist to cope with these potential problems?

10. *Sensitivity.* There is great uncertainty about the future role of energy, including its types, its availability, and its costs. It is important to consider a variety of scenarios and determine what the company's position should be with respect to each possibility. The form should be resilient to many of the possible outcomes.

More specific examples of energy strategies are available from a 1978 survey of *Fortune* 1,000 firms. According to this survey, 90.5% of the firms responding had plans to continue operations during periods of energy shortage, 74.8% had a corporate energy plan, 64.9% had established an energy department within the firm, 62.3% had added more energy-efficient products to the product line, and 60% had a plan for recycling and reusing materials and products. The questions in the questionnaire represent various strategies worth considering. These strategies and their degree of implementation are shown in Table 6.

To gather information for this study, we developed the following questions which we considered the most important for energy strategic planning in a management accounting context:

- Does your firm include planning for its future energy needs as a formal part of its planning process?
- Is your firm equipped to use an alternative energy in case a shortage develops with the energy you are currently using?
- Does your firm have contingency operating work schedules for periods of energy shortages?
- Does your firm have customer allocation plans to be implemented when reduced energy supplies cause product shortages?
- Does your firm treat energy problems differently between U.S. operations and foreign operations?
- Does your firm have investments or joint ventures in energy related industries for the purpose of assuring energy supplies?

Capital Expenditures and the Energy Audit

The process of establishing energy strategies should lead a firm into developing capital expenditure projects related to its energy objectives. Capital expenditure projects should be subjected to the same economic evaluation that determines other items in the capital budget.[2]

TABLE 6
Implementation of Energy Strategies

Strategies	Response in percentages: Our firm has already implemented this strategy	Response in percentages: This strategy will be implemented*
Adoption of a corporate energy plan	74.8	68.5
Reassessment of customer service standards	32.1	14.2
Increasing minimum order sizes	24.8	7.0
Addition of more energy-efficient products to our product line	62.3	30.6
Propose methods to continue operations during periods of energy shortage	90.5	77.2
Prepare working schedules which would minimize the hardships to employees caused by such things as layoffs, reduced working hours, and furloughs	52.5	38.5
Complete shift to more widely available fuels (for example, coal)	47.0	31.9
Adoption of company plan concerning recycling and reuse of materials and products	60.0	47.4
Our company will employ public service advertising to inform our customers of the energy crisis and how we are responding to it	24.8	7.8
The firm's annual report will have a section specifically addressing the energy situation and its effects on our firm	69.0	31.0
Establishment of an energy department within our firm	64.9	28.9
The energy consumption of all our products will be stated on the nameplate, price tag, and in any advertisment	14.8	5.6

*This is shown as a percentage of those firms indicating they had not implemented the strategy.

Source: Lambert and Stock[1]

A useful starting point for identifying potential capital expenditures to meet energy objectives is the energy audit. The basic objectives of an energy audit are to find out (1) how much energy costs, (2) how and where it is used, (3) how much is wasted, (4) how and what internal and external factors cause changes in energy consumption from one period to the next.[3] The energy audit requires a combination of accounting and technical skills: the accountant determines how much energy costs by reviewing invoices and utility bills, and the technical skills vary depending upon the nature of the company.

Some companies are large enough so that they employ engineers with the technical skills required for an audit, while other companies must use outside consultants. A list of energy consultants is published annually by *Energy User News*[4] and is classified as to: Architectural/Engineering/Design Services, Forecasters/Researchers, and Consulting Engineers. The service capabilities of each consultant are listed as shown below:

Facilities siting
Conservation planning
Energy use, procurement and/or supply analysis
Contingency planning
Environmental impact analysis
Technology assessment
Alternative source analysis (biomass, wind, etc.)
Cogeneration analysis
Methods/Investment strategies
Modeling and Analysis
Payback Analysis
Energy audits
Tax credits
Grant programs
Design services
Mechanical/New construction
Mechanical/Retrofit
Combustion/New construction
Combustion/Retrofit
Heat recovery/New construction
Heat recovery/Retrofit
Energy Management Systems/New Construction
Energy Management Systems/Retrofit

Because of the importance of items covered in an energy audit, we included the following questions in the survey:

- Does your firm conduct energy audits? If yes, how often? Who does it?
- Does your firm periodically review contracts and rates for utilities and municipalities?

We also asked questions with respect to capital expenditure projects:

- Has your firm invested in equipment to control the consumption of energy?

- Does your firm adjust future cash flows connected with capital expenditure projects for anticipated increases in energy costs?
- Does your firm identify the quantity of additional energy or energy savings connected with a project?
- Does your firm have investments or joint ventures in energy-related industries for the purpose of assuring energy supplies? (This is also a strategic planning question.)
- Does your firm use a higher rate of return on investment for energy-related projects than the return on investment used for other investment projects?

The last question seemed unusual to some of the firms we visited for the case studies, but the president of a company making process equipment and providing services to energy-intensive industries indicated that many firms might have higher cutoff rates for energy projects. He said, "Most companies set the expected rate of return from energy-saving investments at a level about twice as high as that for mainstream business investments."[5]

Management Control and Energy Management

The management control process for energy should include goals, organization and measurements. The national director of energy consulting for a major public accounting firm says that, "The controller's role in energy reporting is the same as in financial matters; the only difference is the unit of measure. BTUs are used to budget and measure energy instead of dollars." His statement indicates that there is a need for establishing goals in the form of an energy budget against which consumption is measured.[6] The same expert feels energy committees are an essential part of the organization. Energy committees should include the chief executive officer, controller, vice-president of manufacturing, corporate plant engineer, vice-president of marketing, and vice-president of product design. In addition, the energy committee should have a top management representative at each location. The federal government, in its *Energy Conservation Program Guide for Industry and Commerce,* also stressed the need for energy committees.

Dow Chemical was one of the first companies to use management control concepts in developing an "Energy Accounting System."[7] The key to Dow's system is measuring monthly production output in BTUs and the energy value of its inputs. The goal is to operate at maximum energy efficiency. An important part of Dow's system is the organizational structure, which includes a manager of energy conservation, a corporate energy accountant and, in each division, a part-time energy coordinator, an energy accountant and a plant cost accountant responsible for utility data.

Northeast Utilities, a New England electric utility holding company, described an "Energy Management Program" for its commercial and industrial customers which uses sound management concepts that should be present in most programs. They are reproduced in Table 7, which appears on page 20.

The management control process for energy, consisting of goals, organization and measurements, was covered in the mail questionnaire by a variety of questions. Concerning the goals portion, we asked:

- Has your firm specified goals for reducing its energy needs?
- Has your firm established guidelines for controlling energy consumption?

Questions related to the organizational structure for energy management were:

- Does your firm have an executive who is responsible for energy-related matters? If yes, what is the title?
- Does your firm have a corporate-level energy department?
- Has your firm formed energy conservation committees or policy teams?
- Has your firm added energy-related responsibilities to job descriptions?
- Does your firm have an energy conservation manual?

The last question deserves comment. Although some companies have energy conservation manuals, not much has been written about them. Our research team learned of their existence in the investigative phase which preceded development of the questionnaire, but we could find no examples in the literature. By asking the question and then pursuing it during visits to the companies in the case study phase, the research team gained insights which we will share later in this study.

We covered the measurement phase of management control for energy with the following questions:

- Has your firm established guidelines for measuring current energy consumption?
- Does your firm measure the consumption of energy by production process?

TABLE 7
Northeast Utilities Energy Management Program
Organization and Planning

Once top-level management decides to commit the organization to energy conservation, the next steps are to establish an energy committee and to provide that committee with operating guidelines. The operating guidelines should include:

1. Committee organization with subcommittee support as needed to perform all energy conservation planning and implementation functions.
2. Develop a method to account for and record all energy use. Initiate monthly reporting of energy use. Make comparisons with the use from the previous year(s).
3. Plan and perform energy audits. Analyze results to identify cost-effective opportunities for energy conservation.
4. Set a realistic energy conservation goal. As the program becomes fully developed, revise the goal upward, if possible.
5. Enlist employee support and participation. Solicit ideas and provide incentives.
6. Research and develop energy conservation ideas. Employ consulting services, if necessary.
7. Establish energy conservation procedures and monitor their implementation and effectiveness.
8. Plan and conduct a continuing program of activities to stimulate and maintain interest in energy conservation.
9. Through periodic reporting, keep the organization informed of energy conservation progress and management's commitment to it.

A coordinator, whose sole or primary responsibility is to head the committee, should be appointed by and report directly to top-level management. The coordinator should have qualifications that include a technical and financial background, a familiarity with all operations of the organization, and skills in oral and written communication.

Representatives from each department in the organization should be selected by the management to complete the committee staffing. They should report to the committee coordinator. Each department representative should be completely familiar with the technical operation of his or her group.

- Does your firm measure the consumption of energy by piece of equipment?
- Does your firm have energy supply inventory reporting? If yes, how often?
- Does your firm have energy consumption reporting for management? If yes, how often?
- Does your firm know what the total corporate energy cost for the last fiscal year is? If yes, what is the percentage of corporate energy cost to total corporate costs?
- Does your firm know how much a $1 rise in OPEC cost of a barrel of oil will affect the cost of each of its products?
- Does your firm have a standard cost accounting system? If yes, do energy-related expenses have a separate standard? If no, do you have a separate line item in the budget for energy-related expenses?
- Does your firm prepare energy usage trend reports? If yes, how often?
- In analyzing performance of energy-related expenses against budget or standard, are rate variances separated from usage variances?

In addition to questions on the management control process, strategic planning, and capital expenditures, we included a corporate social responsibility type question—"Does your firm sponsor in-house programs on how to conserve energy?"—and the following questions designed to help analyze the effect of geography, industry, and size on a firm's energy program:

- Does your firm have extensive operations in the: Northeast? South? Midwest? West?
- The best description of your firm's industry is: Mining? Manufacturing? Transportation, Communication, Electric, Gas and Sanitary Services? Wholesale and Retail Trade? Services?
- What were the reported sales in your firm's last annual report?

The complete mail questionnaire is presented in Appendix 1. In the next chapter we describe our research methodology and analyze the results of our mail questionnaire in terms of the energy management process and the management accountant's role.

Notes

1. Douglas M. Lambert and James R. Stock, "Organizing and Implementing the Corporate Energy Plan," *MSU Business Topics* (Summer 1979), p. 9. See, also, *California Management Review* (Fall 1979), pp. 5–11.

2. Robert O. Redd, "The Controller and Energy Management," *Management Accounting* (March 1981), p. 50.
3. Harold Roth, "A New Outlet for Energy Audit Data," *Journal of Accountancy* (September 1981), p. 68.
4. *Energy User News* is published weekly and may be obtained from P.O. Box 402, Martinsville, N.J. 08836.
5. G. N. Hatsopoulous, E. P. Gyftopoulos, R. W. Sant, and T. F. Widmer, "Capital Investment to Save Energy," *Harvard Business Review* (March-April 1978), p. 112.
6. Redd, *op. cit.,* p. 48.
7. Morris Gartenberg, "How Dow Accounts for Its Energy Use," *Management Accounting* (March 1980), p. 10–12.

Chapter 3

The Role of Management Accounting in the Energy Management Process: Results from the Mail Survey

We sent the mail questionnaire to the chief executive officers of 1,000 companies. (See Appendix 1.) We selected the recipient firms from the SEC registrants in the 1980 edition of the *Directory of Companies Required to File Annual Reports with the Securities and Exchange Commission.* The sample was stratified into five SEC industrial classifications, and the number of firms selected in each classification was proportionate to the number of firms from that industry to the total in the directory. The industrial classifications and number of firms selected in each classification broke down as follows: mining (120); manufacturing (505); transportation, communication, electric, gas and sanitary services (120); wholesale and retail trade (135); and services (120). After determining the number to be sampled in each classification, we selected a random sample within the classification.

Response to the Mail Survey

The first mailing yielded 173 usable questionnaires. A second mailing resulted in an additional 82 usable questionnaires for a total of 255. Both mailings were accompanied by appropriate cover letters, reproduced in Appendix 2 and Appendix 3. The responses by industry were: mining, 19; manufacturing, 98; transportation, communication, electric, gas, and sanitary service, 38; wholesale and retail trade, 49; services, 50; and one firm not identifying its industry.

The median revenues of those firms responding was $120 million, with a range of from $1 million to $50 billion. Analyzing the responses by region showed that 41% of the corporations had extensive operations in the Northeast, 51% in the South, 52% in the

Midwest, and 40% in the West. The percentages add to more than 100 because a firm may operate in more than one region.

The possibility of biased results, because those who did not respond might have given significantly different answers from those who did, was subjected to a standard statistical test. The test consisted of comparing the differences between the answers from the first mailing and the answers from the second mailing. Using a 10% significance level, only five responses out of 48 were significantly different. (Note that although there are 35 questions, multiple answers give 48 responses to the questionnaire.) Because differences apply to only 10.5% (5/48) of the content of the questionnaire, we concluded that there was no statistical difference between the two mailings and therefore no bias introduced by nonrespondents.

In examining the results of the questionnaire, we attempted to discern patterns by analyzing the responses with respect to size, industry, region and the cost of energy. We separated the responses by size into large companies with revenues above the $120 million median and into small companies with sales between $1 million and $120 million, and then we separated these groups into industries based on the five standard industrial classifications. The analysis by regions posed some problems because many firms had extensive operations in more than one region. To avoid problems of overlapping, the regional analysis was limited to 133 firms that had extensive operations in only one region. The median cost of energy as a percentage of total corporate costs amounted to 3.5% for 113 firms that knew their percentages. Using 3.5% as the dividing line, we compared the responses of firms with significant energy costs, greater than 3.5%, to those with energy costs less than 3.5% of the total.

In analyzing the responses with respect to the four variables—size, cost, industry, and region—there was the possibility of an interaction effect between variables. Interaction effects for size-industry, size-cost, size-region, industry-cost, industry-region, and cost-region were tested statistically. Only size-industry and industry-region were significant at the 5% level. We performed regional analysis on the responses to all questions, but we found little of significance.

The size-industry interaction occurred because most mining companies responding were small and many of the transportation, communication, gas, electric and sanitary service firms (TCGES) were large. The other three industry groups were about evenly divided between large and small firms. This interaction between size and industry means that in the analysis presented in this chapter, one cannot be sure whether the effects observed for the mining industry were because of industry or the size of firms responding, nor can one be sure whether the results for TCGES were because of the industry or the size of the firms responding.

Strategic Planning

In terms of strategic planning, more than half the firms—51.0%—include planning for future energy needs as a formal part of the planning process, and another 6.3% are in the process of developing this plan. (The response to the strategic planning questions are presented in Table 8.) Yet, as we look at positive responses to specific strategies (ability to use an alternative energy source, 41.2%; contingency operating work schedules, 22.8%; customer allocation plans, 18.1%;

TABLE 8
Percentage Reponse to Strategic Planning Questions

	Yes	No	Do Not Know	In the Process of Developing
Does your firm include planning for its future energy needs as a formal part of its planning process?	51.0	41.6	1.1	6.3
Is your firm equipped to use an alternative energy in case a shortage develops with the energy you are currently using?	41.2	53.3	1.2	4.3
Does your firm have contingency operating work schedules for periods of energy shortages?	22.8	68.6	3.5	5.1
Does your firm have customer allocation plans to be implemented when reduced energy supplies cause product shortages?	18.1	72.9	5.5	3.5
Does your firm have investments or joint ventures in the energy-related industries for the purpose of assuring energy supplies?	18.1	79.9	1.6	.4

and investments or joint ventures to assure energy supplies, 18.1%), we see the response rate declining.

There seems to be considerable room for improvement in the area of strategic planning because 41.6% of the firms do not include planning for future energy needs as a formal part of their planning process. An analysis by size, cost and industry shows where strategic planning might be employed effectively. According to the results shown in Table 9, companies with less than 120 million in revenues and companies in mining, wholesale and retail trade, and services are lower than average with respect to including energy needs in their strategic plans. One could argue that if energy is not a significant cost, why be concerned, yet this is not a valid argument given the analysis of the future energy situation presented earlier. None of the firms could operate without energy.

In terms of the other strategic factors—alternative energy sources, contingency operating work schedules, customer allocation plans, and investments to assure energy supplies—in Table 9 we see that the larger firms are superior to the smaller firms, but both could be improved. The same factors are pursued more by companies where energy cost is significant than where it is less so.

Certain strategic factors seem to be more important in some industries than in others. According to answers in Table 9, being able to use an alternative energy source is far more important in manufacturing than in other industries. Understandably, customer allocation plans are more significant for the transportation, communication, electric, gas, and sanitary services segment (TCEGS) than for other industries. The mining industry, which seems to do little planning for energy needs, is one of the classifications with a high percentage of investments in energy-related industries for the purpose of assuring energy supplies. No analysis by regions is presented because the results were not statistically significant.

Capital Expenditures

Eighty percent of the firms in our survey either have invested in equipment to control the consumption of energy (76.9%) or were in the process of doing so (3.1%). For details, see Table 10. A smaller percentage adjust cash flows for anticipated increases in energy costs (56.1%), and an even smaller percentage (48.2%) identify the quantity of additional energy or energy savings connected with a project. We effectively disproved the idea that most firms set the expected rate of return from energy projects higher because only 4% said they do it. As we discussed answers to the question of investments in energy-

TABLE 9

Analysis of Percentage Response "Yes" and "In the Process of Developing" by Contributing Factors for Strategic Planning Questions

	By Size		By Cost		By Industry				
	Larger than Median in Revenues	Smaller than Median in Revenues	Greater than Median Energy Cost/Total Cost	Less than Median Energy Cost/Total Cost	Mining	Manufacturing	TCEGS	Wholesale & Retail Trade	Services
Does your firm include planning for its future energy needs as a formal part of its planning process?	67.5	47.7	74.1	62.5	26.3	66.3	71.1	44.9	52.0
Is your firm equipped to use an alternative energy in case a shortage develops with the energy you are currently using?	56.7	30.3	50.0	48.2	15.8	70.4	47.4	16.3	36.0
Does your firm have contingency operating work schedules for periods of energy shortages?	35.8	21.2	35.2	23.2	10.5	31.6	31.6	20.4	32.0
Does your firm have customer allocation plans to be implemented when reduced energy supplies cause product shortages?	36.0	11.5	27.8	14.3	10.5	16.5	57.9	10.2	20.0
Does your firm have investments or joint ventures in energy-related industries for the purpose of assuring energy supplies?	25.0	12.2	18.9	14.3	42.1	14.3	47.4	4.2	8.0

TABLE 10
Percentage Response to Capital Expenditure Questions

	Yes	No	Do Not Know	In the Process of Developing
Has your firm invested in equipment to control the consumption of energy?	76.9	18.8	1.2	3.1
Does your firm adjust future cash flows connected with capital expenditure projects for anticipated increases in energy costs?	56.1	39.6	3.5	.8
Does your firm identify the quantity of additional energy or energy savings connected with the project?	48.2	47.8	1.6	2.4
Does your firm have investments or joint ventures in energy-related industries for the purpose of assuring energy supplies?	18.1	79.9	1.6	.4
Does your firm use a higher return on investment for energy-related projects than the return on investment used for other investment projects?	4.0	83.6	11.6	.8

related industries to assure supplies in the preceding section, we included them in Table 10 for comparison purposes only.

In the analysis of capital expenditure questions by factors, presented in Table 11, it can be seen that, regardless of size or the percentage of cost that energy represents, most firms have invested in equipment to control the consumption of energy, with the exception of firms in the mining industry. Because adjusting cash flows for anticipated increases in the cost of energy is regarded by some people as a sophisticated technique, it is interesting to see that, regardless of size, the majority of firms make this adjustment. Among the 110 firms which knew the percent of energy cost to total cost, around two thirds of them adjusted future cash flows for anticipated increases in energy costs.

The survey revealed that only about half of all firms quantify the additional energy or energy savings connected with a project—48.2% said yes, and 2.4% were in the process of developing it. As Table 11 shows, the smaller firms are less apt to quantify, as are those where energy is less than 3.5% of corporate cost. The industry analysis shows that firms in mining, 15.8%, wholesale and retail trade, 32.7%,

TABLE 11

Analysis of Percentage Response "Yes" and "In the Process of Developing" by Contributing Factors for Capital Expenditure Questions

	By Size		By Cost		By Industry				
	Larger than Median in Revenues	Smaller than Median in Revenues	Greater than Median Energy Cost/Total Cost	Less than Median Energy Cost/Total Cost	Mining	Manufacturing	TCEGS	Wholesale & Retail Trade	Services
Has your firm invested in equipment to control the consumption of energy?	87.5	73.5	88.9	75.0	42.1	90.8	76.3	81.6	74.0
Does your firm adjust future cash flows connected with capital expenditure projects for anticipated increases in energy costs?	57.5	56.8	66.7	69.6	57.9	63.3	52.6	46.9	56.0
Does your firm identify the quantity of additional energy or energy savings connected with the project?	63.3	38.5	60.4	48.2	15.8	66.0	63.2	32.7	40.8
Does your firm have investments or joint ventures in energy-related industries for the purpose of assuring energy supplies?	25.0	12.2	18.9	14.3	42.1	14.3	47.4	4.2	8.0

and services, 40.8%, are less inclined to quantify energy in their projects than are manufacturing, 66.0%, and TCEGS, 63.2%.

Energy Audits and Rate Reviews

Given the importance of energy audits and their prominence in publications of energy, it was surprising that only 42.5% of the firms said yes to the question, "Does your firm conduct energy audits?" An additional 7.5% were in the process of developing the audits, which may indicate an increasing awareness of this technique. More encouraging was the response to "Does your firm periodically review contracts and rates for utilities and municipalities?", where 65.9% said yes. Details about the answers to both of these questions appear in Table 12.

TABLE 12
Percentage Response to Energy Audit and Rate Review Questions

	Yes	No	Do Not Know	In the Process of Developing
Does your firm conduct energy audits?	42.5	47.2	2.8	7.5
Does your firm periodically review contracts and rates for utilities and municipalities?	65.9	31.7	1.6	.8

Because only 107 firms answered that they conducted energy audits, the number of responses to "how often?" and "who does it?" are smaller than the number of responses analyzed for other questions. The reader should bear this in mind when considering the analysis. One hundred and one firms indicated the frequency of their energy audits as follows: 43 said variable, six daily, seven monthly, seven quarterly, four semiannually, 26 annually, and eight greater than annually. There seems to be no strong discernible pattern here because of the large "variable" and "greater than annually" response. One hundred firms indicated who did their energy audits. We classified the results as own employees, 68%, outside consultants, 23%, and a combination of own employees and outside consultants, 9%. An analysis of responses to energy audit questions by size, cost and industry is presented in Table 13.

TABLE 13
Analysis of Percentage Response
by Contributing Factors for Energy Audit and Rate Review Questions

	By Size		By Cost		By Industry				
	Larger than Median in Revenues	Smaller than Median in Revenues	Greater than Median Energy Cost/Total Cost	Less than Median Energy Cost/Total Cost	Mining	Manufacturing	TCEGS	Wholesale & Retail Trade	Services
Does your firm conduct energy audits? Percent "yes" or "in process of developing."	62.7	38.9	64.8	54.5	5.6	58.2	55.3	49.0	46.9
Who conducts energy audits?									
Percent using: own employees	76.7	53.8	75.9	63.0	100.0	68.9	83.3	55.5	61.1
outside consultants	15.0	35.9	13.8	29.6		20.0	11.1	38.9	27.8
both	8.3	10.3	10.3	7.4		11.1	5.6	5.6	11.1
Does your firm periodically review contracts and rates for utilities and municipalities? Percentage "yes" or "in process of developing."	75.6	59.2	74.0	63.6	31.6	76.5	76.3	63.8	44.0

The larger companies use energy audits more often than the smaller companies—62.7% versus 38.9%. The more significant the energy cost, the more likely an energy audit is for those firms where energy cost exceeds 3.5% of total cost; 64.8% of them have energy audits while only 54.5% of the other groups use them. Only one firm out of 18 in the mining industry responding to the question indicated an energy audit—and it is performed daily.

As may be noted from Table 13, 76.7% of the larger firms use their own employees for energy audits, while only 53.8% of the smaller firms use their own employees with a consequent heavier use of outside consultants. A higher percentage of companies with significant energy costs use their own employees for energy audits—75.9%—than those where the energy cost is less significant—63.0%. Larger firms probably have more inside technical expertise, and those who have significant energy costs have more need for the inside technical expertise necessary to conduct energy audits. This idea holds true when we look at the industry analysis in Table 13. The technically based industries do a higher percentage of their energy audits with their own employees than do wholesale and retail trade, with the latter making more use of outside consultants. Incidentally, because only one mining company performs energy audits, the 100% use of own employees represents only one firm.

Larger firms are more likely to review contracts and rates periodically for utilities and municipalities than are smaller firms as 75.6% of the former versus 59.2% of the latter indicated this practice. The difference with respect to significance of energy cost is not as great, but there is a notable difference by industry. Manufacturing and TCEGS industry segments responded about equally, around 76%, with wholesale and retail trade, 63.8%, services, 44.0%, and mining, 31.6%. If a review of contracts and rates for utilities and municipalities is useful for three-fourths of the firms in manufacturing and TCEGS, it would seem to be useful for all industries, and the other classifications could improve their performance.

Goals for Energy Management

The first part of the energy management control process is to establish goals. In our questionnaire we asked about goals for reducing energy needs, specific recommendations for reducing energy consumption, and guidelines for controlling energy consumption. Before the survey, we felt that a progression from overall goals for reducing energy to specific recommendations was natural and that more firms would have

TABLE 14
Percentage Response to Energy Goals for Energy Management Questions

	Yes	No	Do Not Know	In the Process of Developing
Has your firm specified goals for reducing its energy needs?	51.4	37.5	.8	10.3
Has your firm established specific recommendations for reducing energy consumption?	65.5	25.5	1.2	7.8
Has your firm established guidelines for controlling energy consumption?	45.5	45.5	1.6	7.4

goals than recommendations. Further, we thought that almost all firms would have guidelines for controlling energy consumption whether they desired to reduce energy consumption or not. The results of the three questions we analyzed with respect to the goals and presented in Table 14, however, were exactly the reverse of expectations.

A higher percentage of firms have specific recommendations—65.5%—than have goals that the recommendations are supposed to accomplish—51.4%—and only 45.5% of the firms have guidelines for controlling energy consumption. These figures reveal weaknesses in the management control process which will be analyzed subsequently. It is worth noting that a large number of firms are in the process of developing energy goals for management controls, based on the responses to all three questions.

In analyzing the responses by contributing factor (Table 15), it can be seen that for all three questions, larger companies and companies with more significant energy costs set goals more often than the smaller companies or those with less significant energy costs. It is alarming that only 50% of the companies under $120 million in sales have specified goals for reducing energy needs and that only 44.7% have guidelines for controlling energy consumption.

Even though 19 mining firms replied, their attention to goals is extremely poor on all three questions, compared to the other industries analyzed in Table 15. The regional analysis produced significant results for the question about specific recommendations for reducing energy consumption. Firms in the Northeast and Midwest gave this question a favorable response of 77.1% and 81.1%, respectively, while the West and South gave a 54.5% and 57.1% favorable

TABLE 15
Analysis of Percentage Response "Yes" and "In the Process of Developing" by Contributing Factors for Energy Goals

	By Size		By Cost		By Industry				
	Larger than Median in Revenues	Smaller than Median in Revenues	Greater than Median Energy Cost/Total Cost	Less than Median Energy Cost/Total Cost	Mining	Manufacturing	TCEGS	Wholesale & Retail Trade	Services
Has your firm specified goals for reducing its energy needs?	73.3	50.0	77.7	58.9	15.8	73.5	65.8	54.2	59.2
Has your firm established specific recommendations for reducing energy consumption?	80.8	65.9	81.5	80.4	26.3	82.7	73.7	67.3	78.0
Has your firm established guidelines for controlling energy consumption?	62.5	44.7	70.4	53.6	10.5	55.1	57.9	46.9	66.0

response respectively. It may be that the location of energy supplies causes this difference because the cost of energy as a percent of total cost was not sensitive to this question.

Organization for Energy Management

The questionnaire results show that companies are weak in establishing an organization for energy management. Although 53.7% of the companies have an executive responsible for energy-related matters, only 20.0% have a corporate-level energy department, only 34.6% have formed energy conservation committees or policy teams, only 35.7% have added energy-related responsibilities to job descriptions, and only 19.6% have an energy conservation manual. Details of the responses are given in Table 16.

TABLE 16
Percentage Response to Organization for Energy Management Questions

	Yes	No	Do Not Know	In the Process of Developing
Does your firm have an executive who is responsible for energy-related matters?	53.7	45.5	.4	.4
Does your firm have a corporate-level energy department?	20.0	78.0	.4	1.6
Has your firm formed energy conservation committees or policy teams?	34.6	59.5	1.6	4.3
Has your firm added energy-related responsibilities to job descriptions?	35.7	57.3	3.9	3.1
Does your firm have an energy conservation manual?	19.6	76.1	0	4.3

We analyzed by title the response to a question concerning the title of the executive responsible for energy-related matters, and we attempted to determine if the title represents someone primarily responsible for energy management or someone combining an energy management position with other management duties. The results from 123 firms that gave titles were: vice-president, 34.1%; manager, 26.5%; director, 22.7%; president, 1.5%; and other, 15.2%. Of the same 123

TABLE 17
Analysis of Percentage Responses "Yes" and "In the Process of Developing" by Contributing Factors for Organization

	By Size		By Cost		By Industry				
	Larger than Median in Revenues	Smaller than Median in Revenues	Greater than Median Energy Cost/Total Cost	Less than Median Energy Cost/Total Cost	Mining	Manufacturing	TCEGS	Wholesale & Retail Trade	Services
Does your firm have an executive who is responsible for energy-related matters?	63.3	44.7	63.0	55.4	15.8	65.3	57.9	55.1	40.0
Does your firm have a corporate-level energy department?	35.0	9.1	25.9	28.6	0	27.6	23.7	22.4	16.0
Has your firm formed energy conservation committees or policy teams?	51.7	27.5	51.9	49.0	0	45.9	44.7	37.5	36.0
Has your firm added energy-related responsibilities to job descriptions?	50.8	27.3	51.9	35.7	10.5	41.8	44.7	32.7	46.0
Does your firm have an energy conservation manual?	35.0	13.6	29.6	30.3	0	19.4	39.5	26.5	26.0

firms, 25% indicated that the executive is engaged primarily in energy, and 75% have an energy position combined with other duties.

The analysis by size, cost and industry (Table 17) shows that larger firms have organized for energy management more than smaller firms. While questions such as, "Does your firm have a corporate-level energy department?", are biased toward large firms, the essential question, "Does your firm have an executive who is responsible for energy-related matters?", is important, regardless of size.

Given that 47.7% of the small firms include planning for future energy needs in their planning process (Table 9), 50.0% specify goals for reducing energy needs (Table 15), and 65.9% have specific recommendations for reducing energy consumption (Table 15), it is surprising that only 44.7% (Table 17) have someone responsible for these energy-related matters. It also is surprising that only 63.3% of the companies with revenues in excess of $120 million have an executive responsible for energy-related matters (Table 17).

Answers to the organization questions for the 110 firms that gave their energy cost as a percent of total cost show that the percentage of firms with energy conservation manuals is higher for both categories analyzed—29.6% and 30.3%—than it was for all firms responding, according to results shown in Table 17. Of 255 firms responding to the question concerning energy conservation manuals, there was a subgroup of 110 which also gave a percent of energy cost to total cost. The subgroup's response was 26.4% "yes" and 3.6% "in the process of developing" versus an overall response of 19.6% and 4.3%, respectively, indicating that those who know their energy costs are more likely to have manuals.

The industry analysis in Table 17 shows that the mining industry does little to organize for energy management, and the service industry does not have many firms with executives responsible for energy. In the first three questions analyzed in Table 17, the manufacturing industry is dominant, but it has fewer energy conservation manuals than other industrial classifications except mining.

Measurements for Energy Management

Measurements are the phase of the management control process most directly relevant to the management accountant. Consequently, there were more questions concerning measurements than for some of the other areas. It would seem that the minimal energy management program should answer "yes" to basic questions such as: "Does your firm have energy consumption reporting for management?" "Does your firm know what the total corporate energy cost for the last fiscal year

TABLE 18
Percentage Response to Measurements for Energy Management Questions (Basic Measurements)

	Yes	No	Do Not Know	In the Process of Developing
Does your firm have energy consumption reporting for management?	48.4	47.2	1.2	3.2
Has your firm established guidelines for measuring current energy consumption?	57.1	37.4	1.2	4.3
Does your firm know what the total corporate energy cost for the last fiscal year is?	64.4	30.4	3.2	2.0
Does your firm prepare energy usage trend reports?	37.2	58.9	1.6	2.4

was?" "Does your firm prepare energy usage trend reports?" The "yes" response was 48.4%, 64.4% and 37.2%, respectively, indicating a serious lack of management accounting data. The answers to these and other questions are presented in Table 18.

There are some surprises in the responses. For instance, 57.1% said that their firm has established guidelines for measuring current energy consumption, but only 48.4% have energy consumption reporting for management, and only 37.2% convert these data into energy usage trend reports. Are firms establishing guidelines for measuring but not reporting what they have measured? Why do so few firms track trends in energy usage? Could it be that many firms have energy programs but do not involve their accounting departments?

The frequency of energy consumption reporting is important for management control. As mentioned previously, only 48.4% of the firms indicated they have energy consumption reporting. Within that group, 66.1% report monthly, 12.5% report quarterly, and 12.5% report annually. Only monthly and quarterly consumption reports would be current enough for control. Energy usage trend reports also are useful for management control, but only 37.2% reported these. The frequency of energy usage trend reports given was 50.6% monthly, 18.4% quarterly, and 18% annually. Again, only monthly and quarterly reports would seem current enough for control.

Responses to the question about energy consumption reporting are

TABLE 19

Analysis of Percentage Responses "Yes" and "In the Process of Developing" by Contributing Factors for Measurements (Basic Measurements)

	By Size		By Cost		By Industry				
	Larger than Median in Revenues	Smaller than Median in Revenues	Greater than Median Energy Cost/Total Cost	Less than Median Energy Cost/Total Cost	Mining	Manufacturing	TCEGS	Wholesale & Retail Trade	Services
Does your firm have energy consumption reporting for management?	63.3	41.2	75.9	57.1	15.8	62.9	63.2	36.7	48.0
Has your firm established guidelines for measuring current energy consumption?	71.4	53.0	74.1	57.1	21.1	67.3	73.7	55.1	61.2
Does your firm know what the total corporate energy cost for the last fiscal year is?	65.5	67.2	100.0	98.2	26.3	74.2	57.9	72.9	66.0
Does your firm prepare energy usage trend reports?	57.1	23.7	50.0	43.6	10.5	51.0	42.1	30.6	34.7

analyzed in Table 19 in an attempt to discover why so few firms (48.4%, from Table 18) report this information. As can be seen in Table 19, a higher percentage of larger companies report—63.3%—versus 41.2% for the smaller. Companies with significant energy costs are more likely to report energy consumption than those where it is less important—75.9% versus 57.1%. While this figure is not surprising, the fact that only 75.9% of those companies where energy cost is more than 3.5% of total cost report this energy consumption for management *is* surprising. If energy cost is that important, then the reporting percentage should be higher. The analysis by industry shows that a low percentage of firms in the mining industry and fewer than half in the wholesale and retail trade and service segments report consumption.

The majority of firms have established guidelines for measuring energy consumption (57.1%, Table 18). Large companies and companies with significant energy costs have a higher percentage of guidelines than small companies and companies with lower energy costs, according to data in Table 19. The majority of the firms in each industry segment have established guidelines, except mining with only 21.1%.

In analyzing the response to the question, "Does your firm know the total corporate energy cost for last year?", it is amazing to see that only 64.4% know (Table 18). In Table 19 it can be seen that when considering those who know or who are developing the information, the percentage is about equal for large and small firms—100.0% and 98.2%, respectively. Obviously, if one knows the total energy cost, one can calculate the percentage to total cost, which explains these very high percentages. Looking at the industry analysis, only 26.3% of the mining companies know their energy cost, while close to three-fourths of the manufacturing companies know, and other segments fall between those two.

Energy usage trend reports would seem to be an extremely important tool to track the past, to assist with projections of the future, and to see the effectiveness of energy conservation. Usage has to be tracked separately from cost because of the significant price increases of the post-1973 period, yet only 37.2% of the firms reported that they prepare these reports, according to data in Table 18. Less than one-fourth of the small firms prepare energy usage trend reports, as can be seen in Table 19, versus 57.1% of the larger companies. Even where energy cost is significant, only 50.0% of the firms prepare trend reports. Except for manufacturing, 51.0%, less than 50% of all industry classifications prepare trend reports, with mining lowest at 10.5%.

There were several questions in the measurement section that

TABLE 20
Percentage Reponse to Measurements for Energy Management Questions (Information Characteristics)

	Yes	No	Do Not Know	In Process of Developing	Not Applicable
Does your firm have a standard cost accounting system?	49.8	45.1	3.5	1.6	0
If yes, do energy-related expenses have a separate standard?	29.0	60.3	9.9	.8	0
If no, do you have a separate line item in the budget for energy-related expenses?	47.7	48.4	2.3	.8	.8
In analyzing performance of energy-related expenses against budget or standard, are rate variances separated from usage variances?	34.5	56.3	7.6	.8	.8
Does your firm know how much a $1 rise in the OPEC cost of a barrel of oil will affect the cost of each of its products?	15.0	74.3	5.9	.4	4.4
Does your firm measure the consumption of energy by production process?	31.6	60.9	1.6	1.6	4.3
Does your firm measure the consumption of energy by piece of equipment?	29.0	67.8	1.2	.4	1.6
Does your firm have energy supply inventory reporting?	19.0	76.7	2.7	1.2	.4

related to management accounting information systems characteristics. Answers to these questions are analyzed in Table 20.

Notice that close to 50% of the firms have standard cost accounting systems, but only 34.5% separate rate variances from usage variances. The latter percentage is similar to the 37.2% that prepare energy usage trend reports in Table 18. Evidently it is not a customary practice to do the routine cost accounting analysis of rate, usage and usage trends iñ close to two-thirds of the firms.

Standard cost accounting systems are used more frequently by large companies—63.3%—than small companies—40.9%—and their use is not significantly affected by the importance of energy cost to the firm. See Table 21. Naturally, the manufacturing firms reported extensive use of standard cost accounting systems—79.6%—which is much higher than it was for other classifications, according to data in Table 21.

Among the firms with standard cost accounting systems, only 29.0% have a separate standard for energy-related expenses. See Table 20. The other firms probably include this information in an aggregate "nature of cost" category. Among the firms without standard cost accounting systems, however, 47.7% include energy-related expenses as a separate line item in their budgets. See Table 20.

A higher percentage of large companies separate rate from usage variance than do the smaller companies, but the relative costs of energy do not seem to be a determining factor, according to data in Table 21. In the industry analysis in Table 21, it is interesting to see that the manufacturing classification has more firms separating rate from usage variance, which one would expect, but it is remarkable that less than half, 46.4%, do this basic analytical analysis.

The degree of detail with respect to energy in cost systems can be observed from answers to the questions: "Does your firm measure the consumption of energy by production process?" (31.6% said yes), "Does your firm measure the consumption of energy by piece of equipment?" (29.0% said yes), "Does your firm know how much a $1 rise in the OPEC cost of a barrel of oil will affect the cost of each of its products?" (15.0% said yes, as reported in Table 20). The analysis of contributing factors in Table 21 shows that large firms are more likely to measure consumption of energy by production process and by piece of energy than small firms. Firms where energy is a significant part of the cost also are more likely to measure by production process and piece of equipment than are those whose energy cost is less significant.

The industry analysis in Table 21 shows the mining industry with a higher percentage of firms who know how much a $1 rise in the OPEC cost of a barrel of oil will affect the cost of each of its products than we

TABLE 21
Analysis of Percentage Response "Yes" and "In the Process of Developing" by Contributing Factors for Measurements (Information Characteristics)

	By Size		By Cost		By Industry				
	Larger than Median in Revenues	Smaller than Median in Revenues	Greater than Median Energy Cost/Total Cost	Less than Median Energy Cost/Total Cost	Mining	Manufacturing	TCEGS	Wholesale & Retail Trade	Services
Does your firm have a standard cost accounting system?	63.3	40.9	51.9	55.4	15.8	79.6	50.0	28.6	32.0
In analyzing performance of energy-related expenses against budget or standard, are rate variances separated from usage variances?	50.4	21.7	41.5	40.0	10.5	46.4	38.9	29.8	27.1
Does your firm know how much a $1 rise in the OPEC cost of a barrel of oil will affect the cost of each of its products?	20.8	10.8	17.6	16.7	52.9	10.2	36.4	8.5	8.7
Does your firm measure the consumption of energy by production process?	43.7	23.5	51.0	29.1	31.6	50.5	37.1	15.9	17.4
Does your firm measure the consumption of energy by piece of equipment?	35.0	26.5	50.0	23.6	21.1	29.6	37.8	19.1	36.7
Does your firm have energy supply inventory reporting?	30.3	11.5	34.0	21.4	10.5	24.0	34.2	8.2	18.4

see in other industries. Also, mining ranks relatively high among industries in measuring the consumption of energy by production process. It is rather surprising that only 10.2% of the manufacturing firms know the effect of a $1 rise in the OPEC cost of a barrel of oil and that only 50.5% measure consumption of energy by production process.

The final measurements question, "Does your firm have energy supply inventory reporting?", was answered "yes" by only 19.0% of the firms, as shown in Table 20. The nature of the business probably is more significant for the response to this question than size or significance of energy cost. In manufacturing and TCEGS, a higher percentage of firms indicated energy supply inventory reporting than did the other industry classifications, but neither manufacturing nor TCEGS had a high percentage—24.0% and 34.2%, respectively.

The Role for Management Accounting

The role for management accounting that emerges from analysis of the results of the mail survey must be based on a subjective evaluation of the data. The reader should look at the data and ask such questions as: "If most of the firms in the survey are doing something in a certain area why don't we?" or "Why aren't we doing the same thing as the other firms in the industry?" Even in areas where there was a low response, energy conservation manuals for example, the management accountant should ask, "Would this benefit the organization?" In such a fashion, the role for the individual firm can be defined from this survey.

In terms of an overall role, looking at management accounting in the aggregate, the subjective opinion of the authors is that management accountants should act in the following areas:

1. Encourage the firm to think of energy as a strategic planning opportunity.

   Do not be lulled by the current (1981) availability of energy; the long-term prospects are serious. Only 57.3% of the firms in the survey had or were developing planning for future energy needs as a formal part of planning.

2. Work with management to actively seek capital expenditure projects to control the consumption of energy.

   Some 80% of the firms surveyed have made or are in the process of making these investments.

3. Improve the capital expenditure analysis by reflecting the effects of anticipated increases in energy costs in future cash

flows and by identifying the quantity of additional energy or energy savings connected with the projects.

Only 56.9% of the firms surveyed adjusted or were planning to adjust cash flows, and only 50.6% identified or were planning to identify energy in the projects.

4. Make sure the firm has specified goals for reducing its energy needs.

   The management control process for accomplishing anything starts with a set of goals. Only 61.4% of the firms surveyed had or were establishing goals to reduce energy needs.

5. Encourage the firm to make someone responsible for energy-related matters, and create an organization of committees or teams to carry out energy goals.

   Responsibility has to be assigned for the management control process to be effective. Only 54.1% of the firms surveyed had or were appointing an executive who was responsible for energy-related matters.

6. Improve measurements of energy consumption and cost.

   According to the results of the survey, it was the weakest part of the energy management process. This should be the management accountant's strength, and it is an important role in the energy management process. Only 51.6% of the firms surveyed prepared or were about to prepare energy consumption reporting for management; only 66.4% knew or were developing the total cost of energy for the most recent year; and only 35.3% analyzed or were preparing to analyze performance of energy-related expenses against budget or standard in terms of rate and usage variances.

Chapter 4

Case Studies of Energy Management Programs

The research team developed case studies based on field visits to 12 companies to verify the results from the mail questionnaire and to expand the study beyond the scope of the mail questionnaire. Only 90 firms, or 35.3% of those responding, said they would cooperate in a case study, and then there were many overlaps by size, industry and region in those volunteering. Working within these constraints, we selected for the broadest coverage available and visited the companies shown in Table 22.

The firms selected for case studies had operations in most of the industries identified by the Department of Energy as the major consumers of energy. This representation by Standard Industrial Classification (SIC) code is shown in Table 23.

In addition to industrial firms, the research team interviewed firms in wholesale and retail trade—Allied Supermarkets, Hardee's, J. C. Penney, Pic N'Pay—transportation and communication—Providence and Worcester Railroad, Southern New England Telephone—and services—Connecticut General Insurance.

Some General Conclusions from the Field Trips

As a result of the field trips, the research team learned some aspects of energy management that were beyond the scope of the mail questionnaire. From this experience, we offer the following generalizations.

1. Energy programs are subject to an evolutionary process.

Energy programs exhibit a process of evolution in which there seem to be three separate stages. In stage one, the company improves its housekeeping; the largest consumers of energy within the firm are identified and efforts to achieve efficiency are under-

TABLE 22
Twelve Companies in Case Studies

Company	Headquarters	Business	Revenues
Allied Supermarkets, Inc.	Detroit, Mich.	Retail & Wholesale Food	$600 million
American Motors Corp.	Detroit, Mich.	Automobile Manufacturer	3 billion
Armstrong Rubber Co.	New Haven, Conn.	Tire Manufacturer	400 million
Champion International Corp.	Stamford, Conn.	Forest Products	3 billion
Connecticut General Insurance Corp.	Bloomfield, Conn.	Insurance	5 billion
Hardee's Food Systems, Inc.	Rocky Mount, N.C.	Food Service Chain	300 million
Hercules Inc.	Wilmington, Del.	Chemicals	2 billion
Lydall, Inc.	Manchester, Conn.	Industrial Manufacturing	100 million
J. C. Penney Co., Inc.	New York, N.Y.	Consumer Merchandise	11 billion
Pic 'N Pay Stores, Inc.	Charlotte, N.C.	Retail Footwear	90 million
Providence & Worcester Railroad Co.	Woonsocket, R.I.	Railroad	10 million
Southern New England Telephone Co.	New Haven, Conn.	Telephone Service	700 million

TABLE 23
Standard Industrial Classification Code (SIC)

SIC	Industry	Companies Visited
20	Food and kindred products	Hardee's
22	Textile mill products	Hercules, Lydall
26	Paper and allied products	Champion, Hercules, Lydall
28	Chemical and allied products	Hercules
32	Stone, clay and glass products	Hercules
33	Primary metal industries	Lydall
34	Fabricated metal products	Armstrong Rubber
35	Machinery, except electrical	American Motors
37	Transportation equipment	American Motors

taken. This step provides large savings at low costs. In stage two, the company seeks to recover waste and extends the search for savings to other areas of energy consumption that were ignored in stage one. The firm applies familiar technology, but the investment and the lower potential for savings lead to a higher cost-per-benefit ratio than was present in stage one. Stage three involves technological innovation, which requires the firm to make costly investments and use unfamiliar technology to develop new ways of doing things. At this point the energy management program has been extended into all of the firm's areas of energy consumption.

2. **A major organizational commitment is required to develop an effective energy management program.**

Top management must be committed to an energy management program if it is to be successful. This commitment must extend beyond delegating energy management to a senior staff person. In an effective energy management program, energy performance has to be one of the criteria by which middle managers are judged in order to involve middle managers in promoting and supporting the energy program. Otherwise they view energy management as a problem for the technical staff, whose expertise is vital in a consulting role but who do not really control the operations that consume energy.

3. **Effective energy goals must consider differences in operating characteristics.**

Overall corporate goals sometimes must be modified to account for differences within the corporation caused by different operating characteristics. For example, goals should be modified to take into account the differences between marketing and producing divisions and between similar operations in different geographical regions with different electrical costs.

4. **Energy management is subject to the same profit maximization goals as general management.**

   Energy management is not conducted with the fervor that all energy conservation is beneficial per se. The same rules of cost/benefit and profit maximization apply to energy management as to other areas of operations. This application means that higher energy prices will lead to more energy conservation and that those firms who do not adjust future cash flows for energy savings connected with capital expenditure projects probably are understating the benefits of energy-saving investments.

5. **Many energy programs exist outside the accounting department but would benefit from involving accountants more.**

   In visits to companies, the research team's initial contact was usually with the executive responsible for energy-related matters; none of these was an accountant. When asked about the role of the accounting department in energy management, the executives frequently replied that the problem had been viewed primarily as a technical problem rather than as a management problem.

   As energy programs developed, executives noticed inaccuracies, lack of consistency, incompleteness in the data concerning management, and noncompliance with directives, all of which led to problems in identifying trends and evaluating performance. These problems could be resolved by involvement of the accounting department because of its expertise in gathering, organizing and presenting data and in designing procedures to ensure the accuracy of data. It is suggested that management accountants take the initiative and approach the energy managers if they have not approached the accountants.

Brief Guide to the 12 Cases

Case studies from interviews with 12 companies are presented in the following pages. A brief guide to each case is presented here to give an overall view of the material contained in the cases. Detailed explanations of items mentioned in the brief guide are contained in the cases.

Allied Supermarkets: This energy program has been effective in reducing electricity for lighting and refrigeration. A checklist of procedures to control energy in these two areas is presented in the case.

American Motors Corp.: The organizational focus of this energy program is at the plant management level, and the goal is to reduce energy usage. Monthly measurements are made, and the format for the monthly energy consumption reporting is presented in the case. AMC has developed a sensible variation on the Department of Energy's efficiency calculation which identifies fixed and variable aspects of energy that the official calculation ignores. The Department of Energy's required calculation and AMC's variation are presented in the case.

Armstrong Rubber: The organizational focus of this energy program is the division, which follows the decentralized management organization of the company. The energy goal is to reduce BTUs per pound of finished product, which means that energy measured in other terms has to be converted to BTU equivalents, as discussed in the case.

Champion International: The organizational focus of this energy program is the mill. The company has reached that stage in the evolutionary process of its energy program where further improvements in energy efficiency will have to come from capital investments, as discussed in the case.

Connecticut General: This insurance company has an energy management program focused on building management. The company's efforts, including employee carpooling, have won energy efficiency awards for CG. An interesting contingency plan for operations during possible periods of energy shortages is discussed in the case.

Hardee's: The organizational focus of this energy program is the restaurant, where energy is needed for cooking, heating, air conditioning, hot water, and lighting.

Hercules: The organizational focus of this energy program is the plant, and each plant has an energy goal. Hercules has to report to the Department of Energy, and these reports show that domestic plants used 34% less energy per pound of output in 1980 than in the base year 1972. The company has a strategic plan which involves less dependence on petroleum and greater efficiency in utilizing energy through co-generation of electricity.

Lydall: The organizational focus of the energy program at Lydall follows the plant-division-group organization of the firm. Lydall is in the first stage of the evolutionary process of an energy program, having started the program only a few years ago. Examples of Lydall's energy consumption reports are included in the case. Although smaller than some of the other manufacturing firms studied, Lydall is following the same energy management program as the larger companies.

J. C. Penney: The organizational focus of this energy program is the store. The company set up a formal energy management process and published a booklet, *Energy Management Process,* available to the public. The energy management process described deals with energy consumed in buildings.

Pic 'N Pay: The organizational focus of this energy management program is the store, as the firm operates a chain of retail shoe stores. Pic 'N Pay has determined that its smaller stores are more energy efficient.

Providence and Worcester Railroad: The focus of energy management for this company is in preventive maintenance. The company is conducting an economic analysis of customers, which includes the energy cost factor, for strategic planning.

Southern New England Telephone: This energy management program is the oldest, started in 1971, and most comprehensive in the group studied. It focuses on all aspects of energy consumption within the company and seeks innovations in every area. Results of energy management have been outstanding, and the company received the President's Efficiency Award. In building management and fleet operations the firm is on the frontiers of new techniques.

Allied Supermarkets

Allied Supermarkets is engaged in the retail and wholesale distribution of food and nonfood consumer products. In November 1978, the company filed for protection under Chapter XI of the federal bankruptcy laws and is operating under Chapter XI at the time this case is written, with hopes of coming out of the arrangement in the near future. In 1979 the company operated 84 retail food stores in five states and 31 drugstores in three states and had revenues of $609 million. Headquarters are in Detroit, Michigan. Because refrigeration accounts for 60-65% of the total energy usage at Allied, reducing energy consumed for refrigeration is the major strategic goal of the energy management program.

Allied Supermarkets uses the payback period for analyzing capital expenditures. Energy-related projects have a required minimum payback of two years, which is shorter than for other types of projects. The rationale for the tighter cutoff for energy projects is that the corporation wants to stay away from high technology equipment, which becomes obsolete too soon, and to avoid the high degree of uncertainty associated with such projects. The firm has invested in energy-saving devices such as strip curtains and electronic pressure controls.

Estimates for the change in energy usage by a project under consideration are provided by engineering estimates and machine ratings. Future cash flows are adjusted for anticipated increases in energy costs, and the anticipated increases generally are obtained from the company's utility. Allied Supermarket's major concern in energy management is not conservation but cost avoidance. As long as energy conservation contributes to corporate profits, the firm will continue an energy conservation program.

Energy management began at Allied Supermarkets in 1974. Very little was accomplished from 1974 to 1976, but from 1976 through 1978 an 8.89% reduction in BTU consumption was realized. The program was abandoned in 1978 when the company filed for bankruptcy and re-established in 1979 with a goal of reducing energy usage by an additional 3% because energy costs were significant and their control would improve profits. Actual reduction was 3.49%, achieved primarily by relamping and replacing incandescent lamps with fluorescent.

A corporate energy conservation committee, a subcommittee of the operations committee, develops energy programs, controls expenditures in the energy area, and post-audits any energy projects. The energy conservation committee is composed of the director of construction and maintenance, the operations manager, and the divisions manager. They discuss energy management every two weeks in an informal setting which they feel is effective.

A key member of the energy management team is the store manager, who conducts a walk-through survey of the store, assigns responsibilities for energy conservation to store employees, and teaches them effective energy usage. The store manager sets realistic energy conservation goals for his store consistent with corporate goals and past consumption. One way of obtaining a history of consumption is to ask the utility supplying the energy to provide this information as most utilities keep such historical data.

The firm's basic energy management policies follow:

"1. All Stores will be relamped every 30 months with Cool White Reduced Wattage fluorescent tubes. Relamping shall include: all sales area lighting, back room lighting, and case lighting.

2. Each Division shall set up an Energy Conservation Committee, to be composed of at least one individual from the following areas—Operations, Merchandising, and Engineering.
3. Each Division shall implement a refrigerated case cleaning procedure as follows:
   a) Each line of cases will be pulled, shut down, and completely dismantled twice each year.
   b) Cases shall be hosed down, all coils properly cleaned, drains cleaned, and cases reassembled.
4. Each Division shall develop a Preventive Maintenance Program to include condenser cleaning twice each year.
5. Each Division shall implement a program of Color Coding all electric panel circuit breakers. (To reduce possibility of mistakes)
6. Between the hours of 10 p.m. and 8 a.m., lighting shall be reduced in the following areas:
   a) Conventional Stores—All lights OFF, except Security lights.
   b) Conventional Stores w/night stock crew—Reduce lighting by 75%.
   c) 24-Hour Stores—Reduce lighting by 50%.
   d) All Stores turn OFF ALL: Meat Department Foodlamps
      - Case Lighting
      - Decor Lighting
      - Cooler Lighting
      - Prep Room Lighting
      - Parking Lot and Canopy Lighting
      - Conventional Stores—ON at Dusk, OFF 30 minutes after close
      - 24-Hour Stores—ON at Dusk, OFF at Dawn
7. Heating and Air Conditioning Thermostats.
   a) Heating thermostats should be set at 65°F. and Air Conditioning at 75°F.
   b) All Vestibule heaters and Back Room and Dock Unit Heaters shall be turned OFF at 10 p.m. each day. If the unit heaters cannot be turned OFF, thermostat should be set at the lowest setting.
8. Additional Items.
   a) Turn OFF all Prep Room equipment when not in use. Make sure all Prep Room equipment is OFF prior to leaving the store at night.

b) Do not overstock cases. Maintain all load limits.
c) Refrigerate immediately all incoming meat, produce, dairy, and frozen food products. Do not leave sitting outside of cooler.
d) Do not leave receiving doors open for extended periods of time, except when receiving product. This should include Dock Doors. A trailer backed up to a dock acts as a radiator and pulls heat from a store. Make sure the door is closed when a product is not being loaded or unloaded.

9. Clean condensing unit fins on self-contained equipment every 30 days.
10. Make sure all exterior doors are properly weathersealed."

American Motors Corp.

American Motors Corp. (AMC) is the fourth largest U.S. automobile manufacturer. AMC manufactures, assembles and sells passenger cars, four-wheel-drive automobiles, and Jeep four-wheel-drive utility and recreational vehicles and their parts and accessories in the United States and Canada and other countries. Corporate headquarters is located in Detroit, Michigan. Its automobile lines primarily concentrate on the small economy car market. The firm also assembles and sells tactical trucks, transit buses and postal vehicles. Subsidiary operations include the production of lawn and garden tractors and precision injection plastic parts. Revenues for 1979 from these operations were over $3 billion.

American Motors' operations are so dependent on energy that strategically the company is trying to provide standby fuel sources in case shortages develop. For example, AMC is drilling for its own natural gas (which it calls self-help natural gas). Previously the firm stored propane gas as a standby fuel, but that strategy became too expensive.

Energy audits are conducted by plant personnel and corporate personnel. AMC prefers not to use outside consultants because experience with them has not produced benefits equal to cost, and the company feels that its staff can do as well. Audits are performed when the plant is not producing, during long weekends for example. The plant is examined and areas for improvement are noted. Audits also are conducted on the energy reports filed by the plants.

AMC plans capital expenditures over a five-year period. Any project, energy or otherwise, must have a payback period of one year or less to be acceptable unless there are extra budget dollars. To help plant managers determine the project's energy savings, corporate headquarters provides methods for the plants to follow such as surveys to be undertaken by the plants to assess their energy needs and areas of energy waste. Surveys are conducted specifically for illumination, industrial processes, heating and ventilating, and air conditioning. To help plants estimate future energy costs, a corporate energy control department gets information from AMC's utility, from company historical data, and from a variety of services publishing such estimates, and it shares this with the plants.

At AMC the energy conservation goals are set in BTUs for the divisions but translated into dollars for the corporation. The goals are set by an energy conservation committee. Starting in 1972, the goal was to reduce energy costs 5% annually, but as time went by, the cost of energy rose so rapidly the goal had to be modified. By 1980 the corporate goal was a 3% reduction in energy cost although this represented a reduction of more than 5% in BTUs. Plant goals also are set in terms of usage because of vast regional differences in energy costs. Each plant has the same energy conservation goal.

A corporate energy control department, reporting to the manufacturing engineering department, funnels information to the plant managers so that they can implement the energy program. For example, the corporate energy group periodically distributes the *Energy Conservation Manufacturing Bulletin*, which summarizes for the plant managers any new energy regulations and laws, new energy technology, and other related matters. The energy control department also is a clearing house for energy conservation project ideas. In the Energy Conservation Project Report form (at the end of this case), the plant manager indicates the nature of the conservation project, the data implemented, a brief summary of the methodology employed, annual energy savings, and the dollar value of this savings. The energy control department makes this information available to other plant managers for ideas and prepares an energy conservation status report annually which indicates plant performance.

Each plant prepares a Monthly Energy Consumption Report for AMC's energy division (the form is given at the end of the case). Plant managers are required to compute the total monthly energy usage in million BTUs and to measure the costs expended for energy. In addition to the consumption report, an energy efficiency report is provided. Its purpose is to assess how efficiently the plants have used their energy resources as compared to energy usage in 1972. Energy efficiency reporting is requested quarterly by the Department of Energy (DOE) as part of its focus on the industries that are major consumers of energy. An example of American Motors DOE reporting, called "Energy Performance Report," is presented at the end of the case.

American Motors' calculations of energy efficiency are different from the DOE's recommendations on two important factors. First, AMC feels that its energy costs should be divided into a fixed and a variable component. In the cited report, line (F), base nonproductive energy, according to AMC, is the fixed energy usage in millions of BTUs measured as the amount of energy used when the plant is not producing (for example, on Saturdays and Sundays).

The second AMC modification to DOE calculations is in adjusting energy usage for the square footage of the plant. AMC's rationale is that if the square footage of the plant changes, then the fixed energy costs change. Once these adjustments are made, the plant's energy efficiency can be computed. In the report, line (L) computes the adjusted base energy use so that the current energy use can be compared to the adjusted base. The resulting figure is shown in line (M).

American Motors' method of calculating energy performance can show less accomplishment than following the DOE calculation would. The following hypothetical example, shown on pages 57 and 58, proves this:

AMC Calculation

(A)	Base Energy Use (1972 in BTU)	10,000	
(B)	Current Year Energy Use (1980 in BTU)	9,000	
(C)	Percent Change in Energy Use	(9,000 ÷ 10,000)	10,000 = 10%
(D)	Base Energy % For Production Use	80%	
(E)	Base Energy For Production	8,000	
(F)	Base Nonproductive Energy	2,000	
(G)	Base Production Units — 1972	4,000	
(H)	Current Production Units — 1980	4,200	
(I)	Percent Change in Production	5%	
(J)	Base Energy Adjusted for Production	8,000 (1.05) =	8,400
(K)	Base Energy Adjusted for Floor Space (+2%)	2,000 (1.02) =	2,040
(L)	Adjusted Base Energy	(8,400 + 2,040) =	10,440
(M)	Energy Use Change From Base Year	(9,000 ÷ 10,440)	10,440 = 13.79%

DOE Calculation

$$\frac{9{,}000 - 10{,}000\ (1.05)}{10{,}000\ (1.05)} = 14.29\% \text{ decline in energy use}$$

NOTE: Some companies calculate DOE efficiency in terms of BTUs per unit, but the answer is the same:

10,000 BTU ÷ 4,000 units = 2.5 BTU per unit in 1972
9,000 BTU ÷ 4,200 units = 2.143 BTU per unit in 1980
(2.143 ÷ 2.5) 2.5 = 14.29% decline in energy use.

ENERGY PERFORMANCE REPORT

PLANT: ____________________

YEAR: ____________________

(A) BASE ENERGY USE (1972 in MMBTU) ____________

(B) CURRENT YEAR ENERGY USE (MMBTU) ____________

(C) PERCENT CHANGE IN ENERGY USE
(B − A) / A (100) ____________

(D) BASE ENERGY % FOR PRODUCTION USE ____________

(E) BASE ENERGY (MMBTU) FOR PRODUCTION ____________

(F) BASE NONPRODUCTIVE ENERGY ____________

(G) BASE PRODUCTION UNITS (Quantity) ____________

(H) CURRENT PRODUCTION UNITS (Quantity) ____________

(I) PERCENT CHANGE IN PRODUCTION
(H − G) / G (100) ____________

(J) BASE ENERGY ADJUSTED FOR PRODUCTION E (1 ± I) / 100 ____________

(K) BASE ENERGY ADJUSTED FOR NON-PRODUCTION F (1 + %Ft^2 Change) ____________

(L) ADJUSTED BASE ENERGY
J + K ____________

(M) ENERGY USE CHANGE FROM BASE YEAR
(B − L) / L (100) ____________

NOTE: DECREASE = FAVORABLE decrease in energy use
INCREASE = Increase in energy consumption

Energy Division

American Motors Corp.

ENERGY CONSERVATION PROJECT REPORT FOR:

DATE IMPLEMENTED: ____________________

METHODOLOGY BRIEF:

Annual Energy Conservation Derived (BTU $\times$ 10^6)

Annual Dollar Value of this conservation

AMC PLANT:	
CONTACT:	
PHONE:	

Energy Manual File	
Administrative	0
Electrical	0
Mechanical	0
Process Change	0
HVAC	0
Specification	0

Energy Division
AMC

YEAR ________

MONTHLY ENERGY CONSUMPTION REPORT. PLANT: ________ MONTH: ________

PRODUCTION COUNT: ________ DEGREE DAYS: ________ OTHER: ________

PERCENT ENERGY FOR PRODUCTION: ________ PERCENT ENERGY FOR E.P.A. ________

INSTRUCTION: ENTER ACTUAL BTU CONTENT OF ENERGY FOR CONVERSION FACTOR, WHEN DOCUMENTED ON INVOICE. OTHERWISE, USE D.O.E. CONVERSION FACTOR LISTED.

ENERGY SOURCE	CONSUMPTION MEASURE	CONVERSION FACTOR	MILLION BTU	DOLLAR COST
NATURAL GAS NON-INTERR.	MCF	× 1.02		
NATURAL GAS INTERRUPT.	MCF	× 1.02		
LPG	GALS.	× .0955		
PROPANE	GALS.	× .0916		
DIST. F.O.	GALS.	× .138		
RES. F.O.	GALS.	× .149		

ENERGY SOURCE	CONSUMPTION MEASURE	CONVERSION FACTOR	MILLION BTU	DOLLAR COST
BIT. COAL	TONS	× 22.565		
COKE	TONS	× 26.		
ELECTRICAL	KWH	× .003412		
SELF-HELP NATURAL GAS	MCF	× 1.02		
WASTE	(state)	×		
TOTALS	→	→		

SUBMITTED BY: NAME: ____________________

TITLE: ____________________

PHONE: ____________________

SIGNATURE: ____________________ DATE: ____________

Armstrong Rubber Co.

Armstrong Rubber is engaged in the design, manufacture and distribution of tires and tubes which are sold for use on automobiles, trucks, buses and other vehicles. Tires and tubes are sold under contract to Sears, Roebuck & Co. for distribution under Sears' brands and to other distributors. Significant operations exist in all parts of the country, and corporate headquarters is in New Haven, Connecticut. In 1980, revenues amounted to more than $400 million.

Armstrong Rubber uses electricity, oil, gas and some propane. It spends more dollars for electricity than for any other fuel, but fossil fuel provides more BTUs. A part of corporate strategy has been to equip most plants with a dual fuel capability so they can use either oil or gas.

Capital expenditures are approved by corporate headquarters. The payback period and return-on-investment method are used for capital project evaluation, which includes adjustments for qualitative factors based upon managerial judgments. The same cutoff point is used for energy projects as for other kinds of capital expenditures. All energy-related projects are examined in terms of the additional energy or energy savings connected with the project. Armstrong Rubber does not directly adjust future cash flows for anticipated inflation, preferring to use present costs and assume that inflation will affect cash inflows and cash outflows in the same way. Management generally feels that any adjustments to present costs are guesses, at best, but energy projects are calculated differently because energy costs in the past several years have climbed faster than the general inflation rate. For energy projects, management increases the selling prices of products 4% per year, then applies the same acceptance criteria as for the other projects.

In the area of energy audits, the corporate facilities department has the staff to conduct mini-audits which are done from time to time if the general managers request them. Also, upon request, the in-house staff provides recommendations on how to conserve energy. Because of the technical expertise in the company, Armstrong Rubber seldom uses outside energy consultants.

The company operates with a decentralized management style under which general managers are responsible for the affairs of the division and are given considerable authority to carry out their responsibilities. This set-up means that the general managers have considerable latitude in how they wish to approach energy management, and, rather than gear a corporate policy toward energy management, several divisional policies related to energy exist. For example, the firm has a consolidated goal of reducing BTU consumption 21.5% by 1985, but it allows general managers to set their own annual goals. Once general managers have examined the energy aspects of their divisions, they set divisional goals for reducing their energy needs. This year their goals ranged from a 10% reduction in BTU consumption for one division to an increase for another division because of a process change.

Champion International Corp.

This firm is a major forest products company engaged in the production and sale of building materials, paper and paper packaging products. Products include softwood plywood, hardwood plywood, hardboard, lumber, fine papers pulp, paperboard, corrugated containers, folding cartons, and various paper bags. Sales in 1979 were $3.7 billion, with 49% coming from the building materials segment, 34% from papers and milk cartons, and 17% from paper packaging products. Corporate headquarters is in Stamford, Connecticut.

Strategic planning has been an effective, integral part of Champion management since 1976. As part of the strategic planning process, the corporate planning staff sets guidelines which give economic assumptions and other parameters for divisional managers to use. Included in these guidelines are energy forecasts derived from national data and adjusted for regional differences in energy costs. The latter point is especially important because energy costs vary by location, especially the cost of electricity. As a part of this strategic planning process, goals for reducing energy are set for the corporation, the divisions and each facility. The goals are different for each division and for each facility because of different situations at each site.

The energy management program is designed to save dollars for the corporation, then to save energy. For example, the firm prefers to use wood rather than natural gas because wood is cheaper, even though natural gas is 75% efficient in terms of utilizing BTU content, while wood is only 60% efficient. Wood also is preferred because of environmental considerations. In addition to an emphasis on economic considerations, Champion is striving to meet the federal guidelines for energy conservation and feels that it can meet those guidelines as long as it has access to the capital markets to finance the projects needed to conserve energy.

In general, Champion treats energy problems for U.S. and foreign operations the same. One exception is in Brazil, where the government has mandated two successive 10% energy reductions. It has become difficult to maintain desired operating capacity levels at 90–95% in Brazil with the energy cutback required by the mandated reductions because there is not enough energy to run the plants.

The future energy situation at Champion may not be as critical as at some other firms because it obtains energy internally from bark and other wood residuals. While the degree of corporate self-sufficiency depends on how one defines the computation, Champion's estimate of self-sufficiency is 50%, and the company also has some coal deposits and natural gas. Nevertheless, because Champion is a large consumer of energy, it feels that all employees need to have an energy awareness.

The company is moving toward greater corporate control over the management of energy resources. Corporate strategic goals are to increase energy productivity and to promote fuel switching through better utilization of wood waste. Achieving these goals will require capital expenditures that will give a corporate planning group more control in energy management because all projects over $25,000 require an appropriation request subject to corporate approval.

Capital investment decisions are based on the return on capital employed, the payback period, and the internal rate of return, which is the most frequently used. Energy projects are looked at essentially in the same manner as other projects. Forecast energy savings on a project under consideration are made by the firm's engineering staff, and future energy costs are projected by the corporation's staff. In coming up with such forecasts, Champion looks at the cost of OPEC oil, analyzes political and economic conditions in Saudi Arabia, considers actions of the Organization for Economic Cooperation and Development, and follows exchange rates. For coal, the firm monitors transportation costs, eastern coal vs. western coal, and the environmental picture. For electricity, the corporation examines the fuel mix of the utility, the utility's cost of capital, the relevant utility control agencies, and the types of customers the utility serves. Additionally, for natural gas, the company considers the impact of the Natural Gas Policy Act of 1978 and how the government might change it.

Energy considerations in capital expenditure projects are very important, particularly when Champion analyzes the installation of new mills. As new mills are being considered, energy-efficient techniques are investigated as are costs for energy efficient designs. If an old mill remains in use, the firm considers ways to make it more energy efficient, such as retrofitting of the heating system (retrofitting is the subsequent installation of equipment after initial construction is completed).

Champion has four operating divisions, each given considerable latitude in its management, and this decentralization concept carries over to energy management. Until recently, the only involvement of corporate headquarters was in monitoring energy usage. During a period of time from approximately 1972 to 1980, the divisions executed basic procedures to reduce energy costs, such as adding insulation, carrying out preventive maintenance, and turning out lights. Energy audits played a role in identifying improvements in these areas. Now the company feels that essentially all of the simple things, items with low cost, have been done and any further reductions in energy consumption will require capital expenditures.

One of the divisions has an energy council which meets quarterly to review what has been done and what can be carried out in the future. The council idea soon may be instituted at the corporate level. In addition, a variety of instructional programs are conducted throughout the firm on energy matters. For example, one program provides instructions on the proper way to operate a boiler. The company also is working on internal communications to improve the exchange of ideas on energy. In these and other ways the corporation tries to induce an energy conservation ethic.

Energy consumption reports are prepared quarterly and in greater depth annually. Data are provided overall, by division, and by mill. Champion converts the consumption of various fuel types into millions of BTUs. The company adjusts these BTUs to obtain a relative energy value which accounts for manufacturing different products. The adjusted energy units per ton have been tracked over time, beginning with 1973, and actual performance has been compared with objectives. The emphasis of the consumption reports is on the four major mills which account for the bulk of energy used at Champion.

Connecticut General Insurance Corp.

Connecticut General Insurance Corp. is in the insurance business, sells mutual funds, and has a real estate operation. With revenues of approximately $5 billion, CG employs more than 13,000 people in more than 100 cities in the United States and has corporate headquarters in Bloomfield, Connecticut. Major subsidiaries include Connecticut General Life Insurance Co., which sells a full line of life and health insurance and annuity products, and Aetna Insurance Co., which provides a full line of property and casualty coverages.

The corporation's major use of energy is for building operations and computers. Energy is needed in the building for lighting, for heating, and for air conditioning. It also is needed to power and to cool CG's computer equipment, which is quite extensive. Connecticut General has been so effective in reducing energy consumption that it has won awards from the state of Connecticut and from Northeast Utilities, its local electric utility supplier.

The firm has several strategies for handling future energy problems. If energy shortages develop, the firm will put into effect its building operating contingency plans, which call for operating on a four-day work week. After studying many alternatives, management decided that the four-day work week for all employees was the most cost-effective way to reduce energy needs. The company also maintains dual fuel capability between oil and gas so that if a shortage develops in one fuel, CG can use another. The firm also is planning to install passive solar heating equipment in a building currently under construction and is evaluating the feasibility of installing a solid waste boiler and a heat pump to reclaim waste heat from the data center.

Energy planning is part of the corporate planning process at Connecticut General, which has separate phases for capital planning and operational planning. Planning for energy consumption, setting goals to reduce energy, and related activities are accomplished within these capital planning and operating phases. When the enterprise is considering major capital investment projects, it prepares a feasibility study, often using outside energy consultants. Persons involved in the feasibility study are asked to identify, among other things, the quantity of additional energy or energy savings connected with the project. The corporation also estimates future energy costs for a project based on estimated energy costs prepared by the engineering department with input from the corporate economist and from energy suppliers. For example, in a recently evaluated capital project designed to upgrade existing boiler controls by using a solid-state direct digital control system, the firm estimated the BTUs that the boiler would generate and translated that to a savings of so many gallons of fuel. Future fuel oil costs were assumed to rise by 20% for the first five years and by 15% for ten years thereafter. The cash flows were discounted to the present, and a payback period of 11 years was computed. Normally, a cutoff point of seven years is used, but the corporation decided to accept the project because of the corporate commitment to conserve energy and a feeling that energy costs might increase even more drastically.

The corporation has not conducted its own energy audits because expected costs exceeded expected benefits. It has, however, engaged consulting firms to examine its buildings and computer facilities to evaluate the present operations and to get input on how to conserve energy. New information from this action was minimal, and the results confirmed management's intuitive judgments.

Connecticut General uses energy conservation committees composed of engineering supervisors and managers. Energy-related responsibilities have been added to job descriptions, but only for people in the Engineering Department, and the performance of the engineers is evaluated partially by their performance of those energy-related responsibilities.

A formal energy management program was instituted in 1973, shortly after the Arab oil embargo. The present energy management organization began in 1976. In 1977, the company set a goal to reduce energy usage 25%, measured in BTUs. Actual reduction was 23.5%. At this stage, the corporation believes it has done all the simple things it can to conserve energy; any further reduction will require capital expenditures.

Energy consumption reports are prepared for management. Connecticut General forecasts consumption and unit costs for electricity, natural gas and oil, and then overall consumption is forecast by converting these to BTU equivalents. Actual usage and actual costs are measured against the forecasts. Differences are examined by the engineering managers, and usage variances are separated from rate variances in analyzing the performance of energy costs.

Connecticut General also sponsors in-house programs on how to conserve energy. From time to time it offers its employees a program titled "From Our House to Your House—Home Energy Conservation Tips" in which employees are shown how they can conserve energy (for example, use of fluorescents and installation of insulation in their homes). It also offers employees the opportunity to purchase energy-related items from a CG supplier at a discount. Connecticut General runs a company carpool and a van pool, and 48% of its employees use these pools. For its successful pooling operations, the corporation won the President's Award for Energy Efficiency in Transportation.

Hardee's Food Systems, Inc.

Hardee's Food Systems, Inc., located in Rocky Mount, North Carolina, is one of the largest food service operations in the United States. The company owned 449 restaurants and had licensed 782 others as of October 31, 1979. It operates food processing facilities and a distribution system which provides food products, restaurant supplies and equipment to most of the restaurants in the chain. In the year that was ended October 31, 1979, company revenues totalled $356,357,000. Later financial data are not available because in 1980 Hardee's was purchased by Imasco Limited.

In 1976, recognizing the importance of energy conservation, Hardee's issued an energy conservation manual to its restaurants titled *Hardee's Energy Monitoring Program* (HEMP) which contained specific recommendations for reducing energy consumption. This manual has been the basis for its energy conservation program. Analysis of several Hardee's' restaurants established that energy was consumed as follows: cooking, 60.6%; heating and air conditioning, 20.9%; hot water, 9.5%; lighting, 9%. The key strategy for energy conservation is developing managerial control at the restaurant, which involves controlling equipment usage. Because it is easier to turn on equipment and leave it on all day, the restaurant manager has to be vigilant to control energy consumption. A checklist used by a restaurant manager to manage equipment usage was developed. The checklist, organized by time, on the hour, told when to turn equipment on and off and had blanks for the manager's initials indicating compliance. As part of HEMP, store managers read gas and electric meters at the end of the month and forwarded the information to headquarters.

Within a restaurant, the cooking elements that consume the most energy are the biscuit oven, 9 KW; the griddle, 14.4 KW; the french fryer vats, 17 KW; and the electric chain broiler, 27.5 KW. The first two are used for breakfast and the others for lunch and dinner. The restaurant manager must balance the desire to service customers promptly with a necessity to avoid having the breakfast and lunch equipment on at the same time. Because the electric bill is based on demand kilowatts, the maximum kilowatts used during the month in any one 15-minute period, having everything on at one time will increase the electric bill significantly.

In HEMP, the company recommended load shedders on electric circuit controllers. This device prevents demand from exceeding a predetermined number of kilowatts. The load shedders also contain controls that can cut selected equipment on and off at predetermined times. At the time the manual was issued, installation of sophisticated load shedders was recommended, and they were estimated to have an average payout of 4.2 years. In 1981, Hardee's installed an automatic control system in a Raleigh, North Carolina restaurant that performed all of the functions of the manager's checklist automatically. This system is on a test basis to determine the payout on the investment and the feasibility of automatic control.

Besides managing the equipment better, Hardee's also is seeking more energy-efficient installations. Recognizing that many states are adopting all or part of the American Society of Heating, Refrigeration, Air Conditioning, and Engineering Code (ASHRAE 90-75, "Energy Conservation in New Building Design") which establishes the energy losses that may be allowed in building design, Hardee's decided to incorporate these requirements in new restaurants. In addition, the company is improving the design of its restaurants by using air lock doors, additional insulation, and the like.

Hardee's sales efforts are organized around five geographical marketing areas. Each area is a profit center, and the incentive to conserve energy comes from the effect of energy costs on profits. Each area has its own program with line authority for the program being the responsibility of the director of operations. One director of operations explained that he felt it was necessary to vary the program each year to maintain employee interest. He used monthly posters in comic cartoon-style, memos, and checklists to vary the form of the energy conservation message.

This same director of operations sponsored annual contests with dollar awards to the manager, assistant manager, and district manager of the winning restaurant. Winners of the contest were determined by improvement in the ratio of energy cost per sales dollar. If one restaurant's costs were 3% of sales in one quarter and 2.7% in another quarter, this cost was considered to be a 10% improvement. The director of architecture and engineering felt that it was difficult to set energy conservation goals because of new products and new equipment. As a minimum, Hardee's strives to keep energy consumption from increasing.

Corporate energy-related matters are the responsibility of the vice-president for construction, but the effect of the cost on operations is the responsibility of line managers under the vice-president of operations. Hardee's devotes considerable management time to energy even though energy costs amount only to 3% of total corporate costs.

Energy cost is measured in dollars monthly and related to sales volume and compared to energy costs budgeted for each restaurant.

Energy cost data are reported as a separate expense code on the restaurants' monthly profit and loss statement prepared by the accounting department. Large variances against budgets are analyzed and explained in terms of cost or volume. Trends in these monthly budget variances also are accumulated and analyzed.

Hercules Inc.

Hercules is a worldwide chemical corporation headquartered in Wilmington, Delaware. Its major product groupings include organic chemicals, plastics, water soluble products, and explosives and aerospace. The organic chemicals are derived primarily from resins for use in products such as adhesives, paint and varnish, printing inks, floor tile, floor polishes, and paper chemicals. The plastics group had sales of $696 million in 1980 and is composed of polypropelene and polyethylene plastics which are used for film, fibers, housewares, paper and textiles. The water-soluble products group, which had sales of $521 million in 1980, are used for thickeners and stabilizers in such products as ice cream, toothpaste, sizing, paint, oil well fluids, and the like. The explosives and aerospace group had $390 million dollars in sales for such products as dynamite, detonators, smokeless powder for ammunition, solid propellant for rocket motors, and graphite fibers.

The international nature of Hercules's operations may be seen by the fact that in 1980, 56% of the profits were from operations outside the United States, and sales outside the United States represented 41% of total sales. The company has wholly owned subsidiaries in Austria, the Bahamas, Belgium, Brazil, Canada, Denmark, England, Finland, France, Germany, Italy, Japan, Mexico, the Netherlands, Spain, and Sweden. It has affiliated companies operating in several other countries as well.

If one took the energy consumption of Hercules in 1979, expressed in millions of BTUs so that different energy sources were comparable, and expressed energy by source, electricity would account for 25.3%, fuel oil number six for 35.4%, natural gas for 15.1%, coal for 5.2%, purchase of steam for 3.3%, wood chips for 9.6%, and other for 6.1%. This breakdown illustrates the current dependence of the company on petroleum-based products. Use of wood chips is possible because the firm is using pine trees for resin for the organic products group.

Hercules's energy conservation program is part of a plan to achieve savings in raw material costs by having greater yields through process improvements and by using lower cost raw materials. Hercules's largest use of energy is for making steam, or heat, for the chemical process, although it uses energy for heating buildings, illumination, air-conditioning, and running machines as well. One of Hercules's corporate strategic objectives is to decrease dependence on increasingly expensive oil and gas. During 1980, the plant at Hattiesburg, Mississippi, switched from fossil fuels to wood at a savings of $1.1 million per year, and four plants were converted so that previously wasted process heat could be used to produce steam. In the future, Hercules plans to convert other plants to coal and wood fuel.

Where appropriate, some plants are moving to co-generation, the production of electricity from steam that is needed in the chemical process. In other words, as the plant produces steam, electricity can be co-generated and can increase the efficiency of the energy usage. Electrical companies using BTUs to produce kilowatt hours have a 40% efficiency. Hercules, by using BTUs to produce kilowatt hours and steam, can achieve an 80% efficiency.

With respect to capital expenditures, Hercules explicitly identifies energy costs and savings in capital projects and adjusts the energy cost for anticipated inflation based on forecasts given on a year-by-year basis. The company makes quarterly short-term forecasts by energy type and by plant location. Some of the projects qualify for the additional 10% tax credit for energy savings, and the capital invested for these projects is separated.

Many capital projects have been implemented for the specific purpose of conserving energy. In some cases these involved a change in process. At one plant, Hercules went from a one-stage to a two-stage coagulation process which required less energy consumption. Other projects have been instituted for increasing the efficiency of burners, recovering heat from waste streams, and cutting out waste by insulation and stopping leaks. In the future the company will continue to look at ways of improving the process to select those methods which are more energy efficient.

The purchasing department periodically reviews contracts and rates for the utilities in municipalities with which the company deals. The purchasing executives are within the group called distribution and supply at Hercules. Hercules engineers conduct energy audits, and sometimes the company hires outside consultants to do them. The manager of energy technology says that 90% of what is discovered in an energy audit is known already by the plants, but the energy audit puts it all together and provides an opportunity for the company to examine the overall picture. The other 10% of the audit offers new ideas.

Hercules controls energy by plants, and each plant has a goal against which plant cost and consumption data are compared monthly. The goal is part of a five-year corporate plan with emphasis on the first year. When a plant manager presents the five-year plan to the vice-president of operations, energy cost is included in the plant budget. Each plant develops recommendations for achieving corporate goals, and programs of each plant contribute to the overall corporate goal, one part of which is to reduce the consumption of fuel oil to 25% by 1985 from the 42% at the end of 1980. There is no difference in energy management procedures between U.S. operations and foreign operations; Hercules's energy management program is corporate-wide.

An executive responsible for energy-related matters holds the title "manager of energy technology" and is part of the production technology group. A plant energy coordinator helps the plant manager and the plant committee in energy management. There also is a corporate policy team, which includes a representative from distribution and supply, government affairs, operations, and a corporate vice-president. The policy team examines the influence of government regulations on what the company is doing and determines overall corporate policy in the energy area.

Sometimes the job descriptions of plant managers contain explicit or implicit energy conservation statements. Plant managers have a responsibility to control cost, and certainly energy is an important part of the overall cost of the plant. The key person in energy conservation is the plant manager, and the key incentive is a bonus program based on safety performance, achieving process improvement objectives, management development, cost control, energy index, and other factors, which shows how they are proceeding against the energy plan.

The energy reporting at Hercules is being handled by engineering. In the future it probably will be taken over by accounting because accounting has the necessary control procedures to make sure the information is accurate. Originally the company made conservation the responsibility of engineering. The reporting was described by the government, and because it was measured in terms of BTUs, which needed engineering judgments about the data, information for the conservation program was prepared by engineering. Government regulations, however, require that an audit trail for energy reporting be maintained, and this area is more familiar to accountants. A quarterly energy conservation report prepared for the U.S. government is made available to management to show corporate progress in conservation. This report's format is dictated by the Department of

Energy, who requires that 1972 be used as a base. The U.S. government has requested this report for the major consumers of energy in such industries as paper, steel, chemical, refineries, and the like. The report is prepared for 34 Hercules domestic plants, and although the U.S. government does not require this report for foreign operations, Hercules applies the same DOE format to its foreign operations to assist them in energy management.

For U.S. governmental conservation reporting, the key variable is the amount of energy used in production during the base year 1972. This figure is derived by dividing the BTUs of 1972 by the pounds of production, or some other measure of production, to arrive at a BTU per unit of production. In Hercules's case, the unit of production selected is the pound, so the base year 1972 is expressed in BTUs per pound. The 1980 BTU data also are converted to a BTU per pound, and the two are compared to find the change.

A simple example will illustrate the reporting process. If in 1972 a firm used 60 BTUs to produce six pounds of product, it would equal 10 BTUs per pound. If, subsequently, in 1980 a firm used 49 BTUs to produce seven pounds of product, that would be seven BTUs per pound. The decline from 10 BTUs per pound to seven BTUs per pound is three BTUs per pound. Three divided by 10, the base year, would give a savings of 30%. The reader will notice that it is not the reduction of energy consumption, per se, that is being measured, but, rather, the efficiency rate.

The quarterly energy conservation report by plant shows the quarterly trend in energy consumption since 1972. The company is working on a method of taking seasonality out of the data to provide a seasonally adjusted trend report. According to the quarterly energy conservation report, Hercules's domestic plants used 34% less energy per pound of output in 1980 than they did in the base year 1972 and achieved a 5% improvement over 1979. This savings was equivalent to 3.8 million barrels of oil per year. Based on the price of oil at the time of Hercules's annual report, which was $33.25 per barrel, this figure represented a cost avoidance of $126 million, which is significant considering that its net income for 1980 was $114 million.

Corporate energy cost is significant for Hercules. At the plant level, power is almost double the cost of plant labor, and although Hercules does not have a standard cost accounting system, it does have a separate line item in the budget for energy-related expenses. The plant's budget is not a flexible budget; it is fixed, and changes in plant volume are handled as explanatory items.

The energy bills are paid by the plants, so there had been a cost allocation scheme in use which allocated plant cost to products. The company decided to meter specific areas to aid in the allocation of energy, and, as a side benefit, the plant managers received a computer printout showing use by areas. In one plant this reporting change led to an 8% reduction of energy consumption, an example of the attention-directing aspect of measurement. The manager of energy technology felt the need to measure consumption by responsibility centers to control energy because managers have a tendency to ignore allocations.

If the situation dictates it, the firm measures consumption of energy by production process or sometimes by piece of equipment. On large pieces of equipment, this measurement is necessary to avoid peak usage electrical demand charges. Hercules maintains monthly inventory supply reporting by location and by type of fuel because the investment in energy supplies is substantial.

It is worth noting that Hercules's energy management achievement of using 34% less energy per pound of output in 1980 compared with the base year of 1972 can be regarded as superior performance. The Chemical Manufacturers Association set as a goal for the chemical industry a 30% reduction by 1985, so one can see that Hercules has far exceeded what the chemical industry is calling for.

Executives at Hercules perform community services in the energy area, and during 1980 they performed an energy audit for the Grand Opera House in Wilmington, Delaware. Hercules also regularly conducts seminars for plant managers to describe some of the procedures that have been successful in conserving energy in other plants. Ideas about energy conservation at work and at home are included in the company newsletter, and other ideas originate from company energy audits. Each plant has an energy conservation coordinator, and there is a biannual headquarters meeting of these coordinators.

Lydall, Inc.

Lydall is a manufacturing firm, headquartered in Manchester, Connecticut, with 1979 sales of $109 million. The corporation has four segments: fiber materials, formed products, metal components, and packaging. The fiber materials units had 1979 sales of $39 million. These units produce fiberboard, paperboard and footwear and tend to be among the corporation's most energy intensive. The formed products segment accounted for $40.6 million in sales. Products under this line of business include dry-laid nonwoven materials, elastomeric (synthetic rubber) components, and molded plastic components. Sales of $19 million were contributed by the metal components group. This segment of Lydall manufactures carbon steel balls and steel shot. The packaging group, which had sales of $10 million, is being deemphasized by the company and need not be discussed.

When Lydall began its energy program in May 1980, it decided that the initial strategic step was to measure energy usage. Goals and specific recommendations would be more effective if the company knew what its record had been. The firm found that one plant alone accounted for 20% of the corporation's total energy cost. Information such as this helped Lydall pinpoint the areas most in need of energy management. Lydall measures energy usage in BTU equivalents which then are related to production volume in tons. The strategic rationale for this measure is that Lydall believes that energy usage depends upon the nature of the product being manufactured, and thus different goals and different recommendations are required for different lines of business.

Capital investment projects for energy purposes are treated the same as other projects in that they must meet the same hurdle rate. The company uses an after-tax payback period adjusted for the cost of capital. Any capital action request must specify the estimated savings or increases in energy usage and must indicate how the project will affect the total plant. Future cash flows connected with capital expenditures projects are adjusted for anticipated increases in energy costs, and energy costs currently are inflated by 15–20% per year. These figures are obtained from outside energy consultants' reports.

Lydall also is supporting an inventor who is working on a way to convert waste paper into fuel. Because the company is familiar with the waste paper business, this invention could provide an alternative source for fuel supplies.

Energy audits by outside energy consultants also have been started by the company. The company used outside consultants because it did not have the technical personnel to conduct such an audit. According to company plans, extensive audits will be conducted every two years. At one time the organization reviewed contracts and rates for utilities and municipalities in detail, but the company says that this procedure is not cost effective because more money can be saved in other simpler ways.

The energy goal set by the firm was to reduce BTUs per production volume by 20% over the 1979 base. The 20% reduction rate was chosen because several other companies such as Dow and Union Carbide have used it and Lydall needed an initial goal. If this goal were achieved, it would translate into a savings of about $1 million, and the firm expects to achieve it by the end of 1982. Moreover, management believes that the long-range potential for energy savings is around 40% of the 1979 base by 1985.

Lydall has designated energy managers at the corporate group, division and plant levels. Most of these positions, however, are combined with other managerial tasks. To help make its energy program successful, Lydall has instituted a bonus incentive plan that is partially a function of how much energy a manager saves.

Lydall reports energy consumption to corporate management on a monthly basis. One report, the Energy Consumption Report, breaks down the monthly energy usage by energy type; includes amount of electric use (cost, kilowatt hours, and millions of BTU equivalent); natural gas (cost, millions of cubic feet, and millions of BTU equivalent); fuel oil (cost, gallons, and millions of BTU equivalent); propane, diesel, or gasoline (cost, gallons, and millions of BTU equivalent); and industrial gas (cost, millions of standard cubic feet, and millions of BTU equivalent). The total BTU equivalents used then are divided by the production volume indicator, measured in tons, to get the energy index, which is used as a measure of managerial performance.

A monthly Energy Conservation Status Report also is prepared for top management. (A copy appears at the end of the case.) This report allows top management to determine how well a particular group is reaching its targeted energy index, what capital projects are being started or completed to help conserve energy, what the energy costs are, and the amount of capital expenditures.

Lydall has a standard cost accounting system. Energy-related costs have been taken out of overhead and reclassified separately in cost reports. Some line item expenses are process electricity, occupancy electricity, occupancy fuel oil, process natural gas, and vehicle propane. Management strongly believes that these items are controllable. In analyzing performance of energy-related costs, Lydall separates rate variances from usage variances. Usage variances are analyzed for any trends.

LYDALL, INC.
ENERGY CONSERVATION
Monthly Status Report

Month________________ Year________________

Group________ Division________ Plant________ Location________

I. *Energy Index Status*

1979 Energy Index — Actual ____________________

1980 Energy Index — Actual ____________

1981 Energy Index — Goal ____________

1982 Energy Index — Goal ____________

1981 Year-to-Date Energy Index ____________

Comments on Year-to-Date Results

II. *Energy Conservation Projects*

1. *Projects Started or Completed Last Month*

   Give brief description of major conservation projects actions, anticipated results, date of completion, capital expended, other relevant information.

2. *Additional Projects Planned for 1981*

   Give brief description of major conservation projects to be undertaken in 1981 with anticipated results, date of completion, capital to be expended.

III. *Energy Cost Status*

1979 Average Cost/mmbtu ____________

1980 Average Cost/mmbtu ____________

1981 Year-to-Date Average Cost/mmbtu ____________

1981 Latest Month Cost/mmbtu ____________

Comments: ____________

IV. *Energy Capital Expenditures*

1980 Energy Capital Expenditures $____________

1981 Year-to-Date Energy Capital Expenditures $____________

Est. 1981 Total Energy Capital Expenditures $____________

J.C. Penney Co., Inc.

J.C. Penney Co., Inc., a Delaware corporation headquartered in New York City, is engaged in selling consumer merchandise and services and has sales of more than $11 billion. Merchandise is sold in the United States through Penney stores, Thrift Drug Stores, and by mail order. The company markets apparel, home and automotive products, drug store merchandise, and insurance.

Energy management at J.C. Penney was a part-time effort until an energy task force established in 1978 determined that the formal, full-time management of energy would be cost effective. The company then proceeded to follow a strategy of establishing a formal structure which would help it conserve energy and reduce energy costs. The formal structure and strategy is described in its booklet, *Energy Management Process,* which is available to the public and highly recommended for an elementary look at energy management because it is well written and easy to understand.

The company's energy management process consists of five steps:

(1) Perform an energy survey of equipment and other energy uses.
(2) Identify the appropriate energy conservation measures which can be applied to reduce the consumption of energy uses identified in the survey.
(3) Determine which materials and equipment will be most suitable in implementing these measures and which provide the best cost/benefit ratio.
(4) Develop and implement an energy action plan.
(5) Audit the results.

The J.C. Penney energy survey is a systematic listing of all equipment and other energy uses and actually consists of several surveys: a lighting survey; a heating, ventilation, and air-conditioning survey; an electrical power survey; and a building envelope survey (explained later).

Lighting represents 40–50% of the electrical consumption of a typical store. A lighting survey examines each lighting area as well as the light source and number of lamps, the lamp watts, the light level (actual and recommended), the method of control, the operating time (actual and required), the condition of the lighting fixture, and the watts per square foot. Any recommendations also should be noted.

Heating, ventilation, and air-conditioning represent approximately 40% of the energy consumed by a typical store. As part of the survey, major equipment is examined, such as boilers, unit heaters, baseboard heaters, refrigeration, air handlers, fans, air intake equipment, duct work, and equipment insulation. For each piece of equipment, one should note the nameplate rating, units in use, fuel used, operating time (actual and required), controlled by, thermostat setting (actual and recommended), and the general condition of the equipment. Any proposed action should be noted, whether it is design, maintenance or operation.

Miscellaneous electrical power represents the last 15–20% of the building energy, and analysis in this area covers escalators, elevators, conveyors, computer equipment, compactors, emergency generators, and kitchen equipment. Persons conducting the survey should state any proposed action. The building envelope survey examines the roof, walls, floor, windows, doors, and any other openings for their general condition and available insulation. Proposed actions again are noted.

After the surveys are conducted, the manager should list all possible actions, then list their estimated costs, estimated savings, the date they can be completed, and their priority. As the actions are completed, the date is recorded. After these actions have been taken, an energy audit is begun which consists of measuring energy usage by electricity, gas and fuel and determining the costs. These data are compared to the degree days in the geographical area of the building and to prior year's historic data. Any differences are analyzed and explained.

When energy-related capital expenditures are contemplated, they are analyzed in the same way as other capital projects. J.C. Penney uses net-present-value and internal-rate-of-return approaches to capital budgeting. The company identifies the quantity of additional energy or energy savings connected with the project based on an analysis by its in-house technical people and adjusts future cash flows for anticipated increases in energy costs. Currently it uses 15%, justified as 3–5% above the inflation rate. This adjustment is not exact because different regions in the United States have different energy costs and because the political situation in the Middle East is so volatile.

Accountants play no role in this energy management process. The company used rate consultants who were paid commissions based on their cost savings and was satisfied with their work. Prior to 1978, the internal auditing staff verified the work of the rate consultants. That is now done by the energy task force because the company feels that energy issues are technical and belong to the engineers.

The corporate goal was to reduce energy consumed per square foot by 40% over the base year 1973. The firm seems well on its way to meeting that objective. In fact, by following the energy management program, several stores already have achieved a 50% reduction in energy per square foot at no sacrifice in customer service. In setting goals, the company recognizes that different buildings have different structures and different sizes and the 40% goal might not be appropriate for all facilities. As a consequence, the company now is developing different goals for different stores.

Energy consumption reporting consists of calculating a store's kilowatts per square foot and its costs per square foot. Data also are compiled for each region and each type of store. An automated analysis system that compares consumption data with historical data and with theoretical indices, that compares actual energy to billed energy, and that computes a utilization factor soon will go on-line. This investment will permit an even better control over energy costs.

Pic 'N Pay Stores, Inc.

Pic 'N Pay Stores, Inc. is a centrally controlled, regional specialty retail chain in the self-select footwear and accessories business with headquarters in Charlotte, North Carolina. Pic 'N Pay emphasizes fashion trends and basic merchandise with an orientation toward middle-income families. The company has 500 stores in 14 states in the Southeast, which produced sales of $91 million in 1980. In December 1980, Pic 'N Pay was acquired by Bata Shoes, a Canadian enterprise.

The company started a formal energy management program in 1978 to assure compliance with DOE building temperature restrictions imposed by President Carter's Administration. When it began a formal energy management program, Pic 'N Pay discovered that energy management systems were not available except to large manufacturing concerns. As energy management systems that were appropriate became available late in the 1970s, the company invested in equipment which would manage peak load usage, setback timing for off-hours, and control the temperature level in the stores. The only problem encountered with this strategy was in maintenance of the system; accordingly, Pic 'N Pay is negotiating a maintenance contract.

The company is following a strategy of reducing its store size, partially for energy management reasons. Previously the typical Pic 'N Pay store averaged 4,000 square feet, but new stores average 3,200 square feet. The buildings hold the same merchandise as before by using the space more efficiently, and, although the construction costs of the new buildings are not significantly lower, the energy savings from reducing the space are quite significant.

Capital expenditures are approved by top management using the payback period as the criterion for selection. Energy projects have the same payback period hurdle rate as other projects, although qualitative features also might be considered. The energy management system installed by Pic 'N Pay in 150 of the corporation's 500 stores has a payback period of two years. The system was installed in stores which were the major consumers of energy, measured in dollars, and which had heating and air-conditioning equipment compatible with the system. Eventually the system probably will be installed in all stores.

At present, the company does not identify the quantity of additional energy or energy savings connected with the capital investment projects because it has not completed the survey phase of energy consumption. When it finishes this phase, the corporation expects to consider additional capital expenditures for energy savings. The company does not adjust the costs of energy for anticipated cost increases because it feels that such estimates are too inexact for numerical quantification. Instead, the company uses a subjective evaluation of energy considerations.

At this point, the primary energy goal is to achieve a maximum efficiency in energy usage, subject to maintaining satisfactory customer relations. No percentage of energy reduction has been set as a goal so far because the firm is in the process of surveying how much energy it uses. Even though the survey has not been completed, however, the company estimates that the energy costs at the stores have been reduced by 15–30%. In the future, the corporation intends to obtain energy usage data in BTUs and to set goals for specific energy reduction. It anticipates adjusting its goals to allow for differences in energy cost for different locations.

The organizational focus of energy management is the store manager. The firm attempts to alert managers to energy issues through a variety of communications. Memos are sent to the store managers on energy topics, such as thermostat controls and lighting, to make the store managers more energy conscious.

Energy costs are measured and the results reported to management by plotting monthly energy bills for each store and graphing them over time to observe trends. Top management receives a list of stores with unacceptable performance along with reports by the store managers on the action they intend to take to remedy the situation. Stores with persistent problems receive a visit from a member of the company's development department, who conducts an energy audit. This person notes trouble spots and offers recommendations which are acted upon.

After installing the new energy management systems, Pic 'N Pay asked the utilities servicing them to review and lower their energy rates. The utilities servicing the company give special rates for using energy-efficient equipment, and Pic 'N Pay management feels that its stores can demonstrate that they qualify. At the time of this case, the company expects to receive the lower rates.

Providence and Worcester Railroad Co.

The Providence and Worcester Railroad Co., headquartered in Woonsocket, Rhode Island, is an interstate freight carrier offering no passenger service. The company operates in Rhode Island, Massachusetts, and Connecticut. Operating revenues were $10 million in 1978 and $11 million in 1979.

Most of the energy costs incurred by the corporation are for its locomotives, although energy costs also are incurred for its cars, trucks and buildings. To a large extent, the company believes that it cannot control its energy costs for several reasons:

1. It is relatively small, which means that it does not have the leverage to significantly affect the price paid for fuel.
2. It is a service organization, and it must maintain its service level or lose customers.
3. It is regulated by the Interstate Commerce Commission, which does not always allow it to drop unprofitable lines.
4. Diesel electric locomotives, which consume most of the firm's energy, have no ready substitutes.

The supply of energy is more important to the company than the cost. If the supply of diesel energy were interrupted, the company could not meet its customers' needs, which might lead to their loss. Faced with this situation, the company has adopted a strategy of cultivating good relationships with its energy supplier. This strategy is considered successful because the firm never has had a serious interruption of its energy supply, not even during the 1973–1974 Arab oil embargo.

The firm has found it nearly impossible to conserve energy, so it has undertaken a few strategic steps to help reduce energy costs. First, the company attempts to keep its railroad tracks in good condition. Second, it has attempted to increase the mileage per gallon for its locomotives. Third, the company tries to have a good maintenance program. In effect, these measures permit the locomotives to run with greater efficiency, thus helping reduce energy costs. The company would like to determine how to stop the locomotive when it is not in use, however, especially in the winter. If the motor is not left running, it cools, and the locomotive becomes difficult to start. The potential savings from this step are quite large.

Energy expended for buildings is not significant. Actions to reduce building energy costs are undertaken only when a building is modified. For example, when its computer room was expanded, the company studied how it should provide the cooling and what energy to use. Studies also were undertaken to determine whether or not the corporation could utilize the extra heat during the winter.

Energy consumption reports are provided for management as part of an overall monthly performance analysis. Data are provided periodically on energy usage by locomotive, car and truck. If any vehicle's costs are out of line, they are investigated. The Providence and Worcester Railroad Co. recently engaged an expert to determine whether any customers cost too much in terms of energy or other factors relative to their revenues. It hopes to boost profits from this analysis. Other than this measure, the firm feels that its energy costs are, to a large degree, uncontrollable. A few things are being done, but to a more important extent, the corporation strongly desires to maintain its supply of energy and will do what it can to assure itself of a source.

Southern New England Telephone Co.

The Southern New England Telephone Company is an associated company of the Bell System, providing telephone service to Connecticut towns and cities except Greenwich, Bethlehem, Southbury and Woodbury. Stockholders owning 70% of the common stock live in Connecticut (AT&T owns 21%). SNET has one of the oldest energy programs in the nation, and in July 1980 was honored to receive one of the first President's Energy Efficiency Awards from President Carter at the White House. (SNET was one of three Connecticut companies and one of 25 nationally to receive the award. To date, approximately 160 awards have been presented throughout the nation.) The company had revenues of over $700 million in 1980 and is headquartered in New Haven, Connecticut.

The strategy of SNET's energy management program is unique

because it is part of a larger one concerned with the company's impact on the ecology of the state. The genesis of this program began in 1971 with a company-wide audit conducted by a consultant, which focused on the impact of Southern New England Telephone on its environment. Each year, starting in 1973, the company has issued a report of its committee on ecological policy covering ecology, energy conservation, and scrap management.

The approach to energy management follows the three areas of consumption—transportation, electricity, and building fuels (the area of greatest savings). Consumption in 1972 was 103,500 BOEs (barrel of oil equivalents), and this figure had been cut to 47,300 in 1979, a savings of 54% over 1972.

The core of the building fuel energy strategy for energy conservation is called building energy management and redesign retrofit (BEMARR). Under this program, which is so extensive it has its own manual, referred to as the firm's energy conservation manual, building engineers examine each location to see where and how energy can be saved. This three-phase process—survey, design, and retrofitting—has resulted in a number of energy-saving actions.

As a result of the examination, the BEMARR team eliminated wasteful reheating systems, replaced large boilers used to provide domestic hot water with more efficient electric or solar hot water systems, and adjusted lighting to suit the task performed. Because lower temperatures require smaller boilers, overly large boilers in older buildings were replaced.

Not only is the size of the boiler important, but so is its efficiency. Great care is taken to ensure that burners are adjusted for maximum efficiency. Besides oversized boilers, the company discovered electric motors that were oversized in relation to the volume of air they had to move. Also, ceiling fans were installed in garages with high ceilings. These fans force the hot air back down, and each saves $500 to $1,000 annually.

Besides making sure that all building energy equipment is the correct size and working at the most efficient operating adjustment, the company constantly seeks new ways to heat, air condition, and light buildings. Some of the areas they are currently looking at include modular boilers, solar installation, solar screens, heat pumps, and microcomputer automation.

Modular boilers have been installed in three locations. A modular boiler is built in several units. In a conventional system, the boiler produces steam in its full capacity even though only a portion of the capacity might be required to heat the building comfortably. In the modular boiler, heat is produced in small units according to need. On a mild winter day, for example, two units may be operating, but during a subfreezing spell, all units might be needed. The company also has adopted a strategy of equipping buildings so that they can be heated by natural gas or fuel oil.

SNET has solar installations at ten locations across Connecticut. One of the ten is an experimental photovoltaic system of solar cells which produces electricity. The company also is installing at Bethany a solar furnace partially financed by a grant from the U.S. Department of Energy. The solar furnace has a saucer-like disc with 18 pie-shaped aluminum sections and a collector on metal legs perched atop the saucer. The system enables the collector to follow the sun's path across the sky by computer and automatically return the collector to its starting position at sunset. The solar furnace is designed to provide power for most of the building's telephone switching equipment, provide one-fourth of the heating requirements, and do up to 90% of the work in cooling the building.

The solar screen with which SNET is experimenting is installed in a window and reduces a building's cooling requirements in the summer by stopping solar heat before it reaches the window. In the winter, the screen forms a warm air blanket between it and the window glass, which saves heat.

The company also is experimenting with heat pumps in several buildings. Heat pumps bring in air from outside the building to heat it and, by reversing the process, can throw unwanted heat out-of-doors to cool it.

Microprocessor automation of a building was tested at SNET's Durham equipment office. The microprocessor allows support services people in New Haven, 20 miles away, to monitor temperature, humidity and energy usage. The microprocessor memorizes trends in the building's energy needs that will help in scheduling power.

Microprocessors and sensors control air temperature better than thermostats. A thermostat activates the system when the temperature falls to a preset lower temperature, but by the time the heat is activated, the temperature has dropped lower than the setting. When the air temperature rises to the thermostat high temperature setting, the thermostat will turn the heating device off, but because it was operating at full capacity, the actual temperature will rise above the setting before it starts to cool off. Thus, although a thermostat may be set for 64° and 68°, the actual operating range might be from 62° to 70° with the system working to raise the temperature 8° instead of 4°.

The microprocessor and sensor systems maintain a level temperature. As the cost of energy goes up and the cost of computers is reduced, SNET believes there will be more opportunities to use microprocessors to control energy consumption. The company is so satisfied with the Durham experience that it plans to automate several more buildings.

Perhaps the unique building application that SNET has developed is its data center in New Haven, dedicated in January of 1979. It is a three-story, 110,000-square-foot building, with no furnace. The building is heated by the computers it houses. The circuitry radiates warmth that is trapped in a heat pump and circulated to the office space. In all but the coldest weather there is even an excess of heat which is shared with another SNET building in New Haven.

The data center building is controlled by a computer that routinely monitors and regulates temperature and humidity. In the summer, the building is maintained at a constant temperature by a system using the outside air and water cooled in outdoor towers.

The potential for savings was not as great in the transportation area of energy consumption, yet considerable attention has been focused on this area. SNET has more than 3,000 motor vehicles, the largest private fleet in the state of Connecticut, and in 1978 and 1979 set the record for the most miles per gallon of any fleet in the Bell System. Two-thirds of the vehicles are trucks, so major strategic efforts to reduce weight and size have been undertaken.

Gasoline can be saved by reducing the weight carried. At SNET this means storing equipment for specialized jobs in garages rather than carrying it from place to place on routine jobs. Weight was removed from some vehicles by replacing steel shelving with plastic dividers. Also, the company replaced installation vans with lighter station wagons and larger automobiles with subcompacts.

Other features of the automotive management program that have been effective are: requiring standard transmissions on replacement purchases of automobiles, keeping the entire fleet checked and tuned for high MPG yield, an improved dispatch program by combining installation and repair assignments, strict adherence to speed limits, and use of air foils on supply tractor trailers to reduce wind resistance. The company is experimenting with diesel vehicles and electric cars as potential energy savers and is testing 100% denatured alcohol in passenger cars.

With respect to the electricity the company consumes, the energy program has benefitted from access to Bell Labs basic research in energy conservation while improving communications. Introduction of transistors, diodes, integrated circuits and other solid-state devices has saved energy because these devices require less energy to operate.

SNET identifies the quantity of additional energy or savings involved in capital expenditure projects and adjusts future cash flows for anticipated increases in energy costs. We studied a company project to observe the energy aspects of capital expenditures analysis. The project reviewed was a microcomputer-based building automation system for five buildings. The results were based on a pilot study of a similar system in one building. Cost of the installation was estimated at $99,000. Two buildings were heated with oil and three with electricity. The test building was heated with oil, but its results were projected to the electrically heated building. Based on current energy costs, the project had a $13,624 annual savings and an eight-year economic life. This figure would give an internal rate of return of 2%—not very attractive—but SNET considered the impact of future energy costs and adjusted the cash flows accordingly.

When the company calculated the microcomputer savings using a 15% annual increase in energy costs, it found the microcomputer savings had an acceptable internal rate of return and a positive net present value. SNET performed a sensitivity analysis and discovered that at an annual energy cost increase of 10% the project was not attractive. The difference between acceptable and not acceptable was calculated at a 13% annual energy cost increase. Because management felt the 15% would be more likely, the project was approved.

An interesting sidelight to the project was that savings from one of the buildings did not have a positive net present value, and one other building was marginal, yet the company accepted all five. Justification for installing the microcomputers in all five locations was that they were characteristic of different SNET buildings and would provide the information needed to analyze future installations of such systems in SNET buildings.

The company periodically reviews contracts and rates with utilities and municipalities. The internal auditors check to make sure the proper rate is used, paying particular attention to peak demand, a significant factor in determining rates. Bills are audited monthly. Southern New England Telephone conducts its own annual energy audits of buildings using the company employees as part of the BEMARR program. A copy of the building inspection form is included at the end of this case.

The company has established goals for reducing energy needs as follows: Overall, the goal is zero growth in energy BOEs per thousand customer lines. The overall goal is distributed among various subgoals such as miles per gallon for the fleet and similar measurements. The goals are tied in with specific recommendations and are part of the management objectives in a management by objectives program used as a basis for calculating bonus compensation. Also, energy-related responsibilities are part of job descriptions. Planning for future energy needs is contained in a six-year plan, the first three years of which are in more detail than the others. The plan includes estimated growth and goals to arrive at consumption. Outside consultants estimate future prices.

The energy program is one of the areas of responsibility of the chairman of the committee on ecological policy, who also has the corporate title "general manager—support services." During the course of this study, it was noted that energy duties are part of a general manager's responsibilities. The responsibilities of the committee on ecological policy include reviewing ecological considerations of the following operations: transportation, building, design, materials use and disposition, telephone switching operations, public and employee awareness, building management/vehicular and employee involvement.

There are two important features of energy management measurements: use of a common denominator to aggregate physical consumption—the BOE and use of a measure of volume to observe the physical growth of the operation—thousand-customer lines. Using these measurements to evaluate results, the SNET energy management program has been impressive. In terms of barrels-of-oil equivalents (BOE), the common denominator for aggregating transportation fuels, electricity, and building fuels, consumption has declined from 249,000 in 1972 to 184,000 in 1979, while the volume of business has increased from 1,180 thousand-customer lines in 1972 to 1,468 thousand-customer lines in 1979. Had the company not reduced consumption by 65,000 BOEs, energy cost in 1979 would have been over $3 million greater.

The current method of measuring energy consumption includes calculating an energy-to-building-space ratio (BOE/000 FT^2). For this calculation the company measures energy by building. Energy also is measured by individual vehicle for management control. SNET writes monthly reports for energy supplies and consumption. These reports show consumption for the current period, year-to-date, and a 12-month cumulative by building and type of energy: electricity, fuel oil, and natural gas.

Besides conserving energy in the operation of the company, SNET encourages its employees and the public to save energy. The company promotes ride-sharing for its commuters and provides reserved parking for carpools. The company also features 14 employee-chartered superbuses and 24 van pools. Company publications stress the advantages of pooling, as can be seen from the following quotes: "We figure that sharing a ride with a neighbor can save the average employee up to $700 a year, and that is downright neighborly."

"Ride-sharing is the single most dramatic step the average person can take to combat energy waste. Each commuter car left at home eliminates 250 pounds of pollutants from the air and saves 600 gallons of gasoline per year." SNET also sponsors in-house programs to make its employees energy conscious and informs employees of energy-saving ideas for the home in the company magazine, *Telephone.*

Turn to page 90 for an example of the company's "Building Energy Conservation Inspection" report.

Building Energy Conservation Inspection

Date ______________________

Company ______________________

District ______________________

Bldg. ______________________

City ______________ State ______________

Bldg. □□ ______________ Grnds. □□ ______________

No. Fls. ______________ Cl. Grp. ______________

Yr. Const. ______________ Add'ns. ______________

Pop. Per (000) □□ ______________________

Occupancy—(See HS & BM Cost Results Plan)

Swbd. ______________ % Equip. ______________ %

Off ______________ % Supply Center Garage ______________ %

Q.R. — Quality Rating } Note: Use Tenths
Q.V. — Quality Value }
W.F. — Weighting Factor
Q.I.F. Quality Index Factor
(Determined By Adding W.F.'s Of
All Classifications Rated and Dividing
By 10)

Example — If No Elevators Or Computer Facilities — Q.I.F. = .90

Note any Unusual Conditions that Contribute to a Low Quality Value

Sheet ________ Of ________ Sheets

H	O	L	U
Higher Than Objective	**Objective Band**	**Lower Than Objective**	**Unsatisfactory Band**
	10.0 – 9	8.9 – 7.0	6.9 – 0

	Classification	**Q.R.**	**W F**	**Q.V.**
1	Ventilation		2.0	
2	Heating and Cooling		2.0	
3	Infiltration, Transmission		1.5	
4	Lighting		1.5	
5	Domestic Hot and Cold Water		.5	
6	Elevators, Escalators		.5	
7	Electric Power Systems		.5	
8	Computer Facilities		.5	
9	Conditioned Space Utilization		.5	
10	Controls		.5	

Quality Index Factor = ______

Total Quality Value

Quality Index	
Numerical	**Band**

TQV ÷ QIF =

Note: Round to Nearest Whole Number

Chapter 5

Conclusion

In the first chapter, we asked a question: "Is there a need for management to plan, organize and control energy needs?" A review of the energy situation showed there was a need. Also, analysis of mail response to a question, "Has your firm specified goals for reducing its energy needs?", showed 51.4% "yes" and 10.3% "in the process of developing." Therefore, 61.7% of the companies are paying particular attention to the energy situation and have goals for reducing energy needs.

In defining the role of management accounting in the energy management process—particularly as it relates to capital expenditures, energy audits, and measuring energy costs—we observed the following statistics for firms responding to a mail survey:

- 48.2% identify the quantity of additional energy or energy savings connected with projects
- 56.1% adjust for future cash flows connected with capital expenditure projects for anticipated increases in energy costs
- 42.5% conduct energy audits
- 48.4% have energy consumption reporting for management
- 37.2% prepare energy usage trend reports
- 57.1% have guidelines for measuring current energy consumption
- 34.5% separate the rate variances from the usage variances in analyzing performance against standard or budget

These responses show a need for management accountants to take a more active role in the energy management process.

One possible explanation for the observed absence of some basic management accounting features in energy management programs is that the firms may view energy as a technical problem rather than as a management problem. In the course of a dozen field visits, observing energy management programs, we did not meet an accountant, although we identified ourselves as accounting professors doing a study for the National Association of Accountants. The energy managers usually were technical people. While a company such as Hardee's uses energy cost data in evaluating energy management, the accounting people are not involved beyond supplying the routine cost data. Regardless of the reason for why management accounting is not more involved, the response shows that it *should be.*

If management accounting and management accountants are to take a more active role in energy management, it will be through the efforts of accountants coming forward and making suggestions to management. The following list of questions and responses could be used to grade an energy management program, an initial step in assessing the need for improvement. The list contains the ten most popular components of an energy program, according to responses from the mail survey.

	% Responding Yes
1. Has your firm invested in equipment to control the consumption of energy?	76.86%
2. Does your firm periodically review contracts and rates for utilities and municipalities?	65.87%
3. Has your firm established specific recommendations for reducing energy consumption?	65.49%
4. Does your firm know what the total corporate energy cost for the last fiscal year is?	64.43%
5. Has your firm established guidelines for measuring current energy consumption?	57.09%
6. Does your firm adjust future cash flows connected with capital expenditure projects for anticipated increases in energy costs?	56.08%
7. Does your firm have an executive who is responsible for energy-related matters?	53.73%
8. Has your firm specified goals for reducing its energy needs?	51.38%
9. Does your firm include planning for its future energy needs as a formal part of its planning process?	50.98%
10. Does your firm sponsor in-house programs on how to conserve energy?	50.20%

Even in the preceding list of questions, which had the highest response rate, only four have positive responses over 60%. Considering the nature of the questions, the positive response should have been much higher. This study has shown that there is considerable room for improvement. The management accountant can use the study to take a more active role in energy management. The benefits of better energy management, given the long-term problems with the energy situation described in the Executive Summary, are important for the individual company and the nation.

Appendix 1

The Energy Questionnaire

SURVEY OF MANAGEMENT ACCOUNTING INFORMATION UTILIZATION IN CAPITAL BUDGETING FOR ALTERNATIVE ENERGY SOURCES AND FOR MONITORING AND REDUCING ENERGY CONSUMPTION

Sponsored by the National Association of Accountants

INSTRUCTIONS: FOR EACH QUESTION, PLEASE CHECK THE APPROPRIATE BOX. A FEW QUESTIONS REQUIRE SHORT ANSWERS IN THE SPACE PROVIDED.

	Yes	No	Do Not Know	In the Process of Developing
1. Has your firm specified goals for reducing its energy needs?	☐	☐	☐	☐
2. Has your firm established specific recommendations for reducing energy consumption?	☐	☐	☐	☐

	Yes	No	Do Not Know	In the Process of Developing
3. Does your firm include planning for its future energy needs as a formal part of its planning process?	☐	☐	☐	☐
4. Is your firm equipped to use an alternative energy in case a shortage develops with the energy you are currently using?	☐	☐	☐	☐
5. Does your firm have contingency operating work schedules for periods of energy shortages?	☐	☐	☐	☐
6. Does your firm have customer allocation plans to be implemented when reduced energy supplies cause product shortages?	☐	☐	☐	☐
7. Does your firm treat energy problems differently between U.S. operations and foreign operations?	☐	☐	☐	☐
No foreign operations. ☐				
8. Does your firm have investments or joint ventures in energy related industries for the purpose of assuring energy supplies?	☐	☐	☐	☐

	Yes	No	Do Not Know	In the Process of Developing
9. Does your firm sponsor in-house programs on how to conserve energy?	☐	☐	☐	☐
10. Does your firm have an executive who is responsible for energy related matters?	☐	☐	☐	☐
If yes, what is the title? ______________________				
11. Does your firm have a corporate-level energy department?	☐	☐	☐	☐
12. Has your firm formed energy conservation committees or policy teams?	☐	☐	☐	☐
13. Has your firm added energy related responsibilities to job descriptions?	☐	☐	☐	☐
14. Does your firm have an energy conservation manual?	☐	☐	☐	☐
15. Does your firm identify the quantity of additional energy or energy savings connected with the project?	☐	☐	☐	☐
16. Does your firm adjust future cash flows connected with capital expenditure projects for anticipated increases in energy costs?	☐	☐	☐	☐

	Yes	No	Do Not Know	In the Process of Developing
17. Has your firm invested in equiptment to control the consumption of energy?	☐	☐	☐	☐
18. Has your firm established guidelines for measuring current energy consumption?	☐	☐	☐	☐
19. Does your firm measure the consumption of energy by production process?	☐	☐	☐	☐
20. Does your firm measure the consumption of energy by piece of equipment?	☐	☐	☐	☐
21. Does your firm have energy supply inventory reporting?	☐	☐	☐	☐
If yes, how often? ______				
22. Does your firm have energy consumption reporting for management?	☐	☐	☐	☐
If yes, how often? ______				
23. Does your firm know what the total corporate energy cost for the last fiscal year is?	☐	☐	☐	☐
If yes, what is the percentage of corporate energy cost to total corporate costs? ______				

	Question	Yes	No	Do Not Know	In the Process of Developing
24.	Does your firm know how much a $1 rise in the OPEC cost of a barrel of oil will affect the cost of each of its products?	☐	☐	☐	☐
25.	Has your firm established guidelines for controlling energy consumption?	☐	☐	☐	☐
26.	Does your firm have a standard cost accounting system?	☐	☐	☐	☐
	If yes, do energy related expenses have a separate standard?	☐	☐	☐	☐
	If no, do you have a separate line item in the budget for energy related expenses?	☐	☐	☐	☐
27.	Does your firm prepare energy useage trend reports?	☐	☐	☐	☐
	If yes, how often? ______________________				
28.	In analyzing performance of energy related expenses against budget or standard, are rate variances separated from useage variances?	☐	☐	☐	☐

	Yes	No	Do Not Know	In the Process of Developing
29. Does your firm periodically review contracts and rates for utilities and municipalities?	☐	☐	☐	☐
30. Does your firm conduct energy audits?	☐	☐	☐	☐

If yes,

How often? ____________________

Who does it? ____________________

	Yes	No	Do Not Know	In the Process of Developing
31. Does your firm use a higher return on investment for energy-related projects than the return on investment used for other investment projects?	☐	☐	☐	☐

If yes, approximately how much higher is the return on investment (e.g., 50% more)? ____________________

32. Does your firm have extensive operations in the:

	Yes	No
Northeast?	☐	☐
South?	☐	☐
Midwest?	☐	☐
West?	☐	☐

33. The best description of your firm's industry is:
    - Mining ☐
    - Manufacturing ☐
    - Transportation, Communication, Electric, Gas and Sanitary Services ☐
    - Wholesale and Retail Trade ☐
    - Services ☐

34. What were the reported sales in your firm's last annual report? ______________________

35. Would your firm cooperate with the NAA in a case study of your energy conservation system? Yes ☐ No ☐

NAME OF FIRM ________________________________

NAME OF PERSON WHO ANSWERED THE QUESTIONNAIRE ________________

ADDRESS ________________________________

____

TELEPHONE (   ) ________________

IF YOU HAVE ANY QUESTIONS ABOUT THE QUESTIONNAIRE, YOU MAY CALL HAROLD E. WYMAN (203) 486-3018 OR J. EDWARD KETZ (203) 486-2892.

IF YOU HAVE ANY ADDITIONAL COMMENTS REGARDING YOUR COMPANY'S UTILIZATION OF MANAGEMENT ACCOUNTING INFORMATION IN CAPITAL BUDGETING FOR ALTERNATIVE ENERGY SOURCES AND ITS MONITORING AND REDUCING ENERGY CONSUMPTION, PLACE THEM ON SEPARATE SHEETS. WE WILL REMAIN ANONYMOUS IN OUR ANALYSIS.

THANK YOU FOR YOUR PARTICIPATION.

A STAMPED ENVELOPE WITH RETURN ADDRESS IS ATTACHED FOR YOUR CONVENIENCE OR RETURN TO:

HAROLD E. WYMAN
J. EDWARD KETZ
ENERGY PROJECT
DEPARTMENT OF ACCOUNTING
SCHOOL OF BUSINESS ADMINISTRATION
UNIVERSITY OF CONNECTICUT
STORRS, CT. 06268

Appendix 2

Cover Letter for First Mailing

National Association of Accountants

919 Third Avenue
New York, N.Y. 10022
(212) 754-9700

John Doe Enterprises
Fifth Avenue
New York, N.Y. 00000

Dear Mr. Doe:

Enclosed is a questionnaire designed to gather information for a National Association of Accountants research study on the "Utilization of Management Accounting Information for Capital Budgeting for Alternative Energy Sources and Monitoring and Reducing Energy Consumption." We feel that the world energy outlook makes this a very important project.

The study is being conducted by Dr. Harold E. Wyman and Dr. J. Edward Ketz of the University of Connecticut, and the questionnaire should be returned to them. All information supplied by respondents will be treated confidentially and published results will not identify how any individual company answered. Respondents will be provided copies of the summary of tabulated responses.

Would you please see that the questionnaire is directed to the executive in your organization most able to respond. Thank you for your cooperation in this.

Sincerely,

Stephen Landekich
Research Director

Appendix 3

Cover Letter for Second Mailing

National Association of Accountants

919 Third Avenue
New York, N.Y. 10022
(212) 754-9700

John Doe Enterprises
Fifth Avenue
New York, N.Y. 00000

Dear Mr. Doe:

Previously we sent your firm a questionnaire designed to gather information for a National Association of Accountants research study on energy usage and energy reporting. We have not received your response, so we are sending this follow-up. We feel that the world energy outlook makes this a very important practical project and that most firms will want to participate.

All information supplied by respondents will be treated confidentially and published results will not identify how any individual company answered. Respondents will be provided copies of a summary of tabulated responses.

Would you please see that the questionnaire is directed to the executive in your organization most able to respond? If you have already returned the questionnaire, we hope you will excuse us for this follow-up. If you have not returned the questionnaire, we urge you to fill it out so that we may have a better understanding of how corporations are responding to the energy situation.

The study is being conducted by Dr. Harold E. Wyman and Dr. J. Edward Ketz of the University of Connecticut and the questionnaire should be returned to them.

Thank you for your cooperation.

Sincerely,

Stephen Landekich
Research Director

Bibliography

American Institute of Certified Public Accountants. *Energy Conservation Studies Including Energy Audits* (New York, AICPA:1977).

Anderson, Gary J., Gordon B. Harwood and Roger H. Hermanson. "Energy Audits," *The CPA Journal* (January 1979):31–34.

Belding, John A., "Alternatives to Oil and Gas Through Energy Management," *Journal of Energy* (January-February 1978):2–7.

Boyles, Jesse V. III. "Energy Accounting," *Management Accounting* (February 1979):35–41.

Common, Michael. "The Economics of Energy Analysis Reconsidered," *Energy Policy* (June 1976):158–165.

Commoner, Barry. *The Politics of Energy* (New York, Knopf: 1979).

Department of Energy. *Industrial Energy Efficiency Program,* Annual Report, July 1977 through December 1978 (Washington, D.C.: U.S. Government Printing Office, 1979).

"Energy '79," *Institutional Investor* (July 1979):113–156.

Gartenberg, Morris. "How Dow Accounts for Its Energy Use," *Management Accounting* (March 1980):10–12.

Griffin, James M. and Henry B. Steele. *Energy Economics and Policy* (New York, Academic Press: 1980).

Hatsopoulos, G.N., E.P. Gyftopoulos, R.W. Sant, and T.F. Widmer. "Capital Investment to Save Energy," *Harvard Business Review* (March-April 1978):111–122.

Henry, John P., Jr., V. Eugene Harless, and Jay B. Kopelman. "World Energy: A Manageable Dilemma," *Harvard Business Review* (May-June 1979):150–161.

Hora, Michael E. "Getting On Top of the Firm's Energy Situation," *Business Horizons* (February 1975):30–34.

"How to Make an Energy Audit," *Electrical World* (March 15, 1977):114–115.

"Industry Can Save Energy Without Stunting Its Growth," *Fortune* (May 1977): 186–200.

John, James H. "The Energy Audit Trap," *The Internal Auditor* (February 1980):16–19.

Jorgenson, Dale W. "The Role of Energy in the U.S. Economy," *National Tax Journal* (September 1978):209–219.

Katugampola, Bernard. "The Energy Audit," *The Internal Auditor* (October 1978):93–95.

Lambert, Douglas M. and James R. Stock. "Organizing and Implementing the Corporate Energy Plan," *MSU Business Topics* (Summer 1979):5–15.

Lindsay, Franklin A. "Financing High-Cost, High-Risk Development," *Harvard Business Review* (November-December 1978):161–171.

Lucado, William E. "The Energy Situation: Implications for Strategic Planning," *Business Horizons* (April 1975):26–30.

Manne, Alan S. "What Happens When Our Oil and Gas Run Out?" *Harvard Business Review* (July-August 1975):123–134.

McClure, Charles E., Jr. "Economic Constraints on State and Local Taxation on Energy Sources," *National Tax Journal* (September 1978):257–262.

Moore, Cecil H. and Damon Button. "Auditing Implications of DOE Controls," *The Journal of Accountancy* (August 1979):78–83.

Piper, James E. "Auditing the Power Your Office Consumes," *Administrative Management* (July 1979):22–23, 66.

Redd, Robert O. "Energy Audits and Conservation," *Michigan CPA* (September-October 1977):24–25.

Sant, Roger W. "Coming Markets for Energy Services," *Harvard Business Review* (May-June 1980):6–24.

Scudiere, Paul M. "Justifying Proposals to Save Energy," *Management Accounting* (March 1980):42–49.

Simich, Stevan and Robert Strauss. "The Energy Audit," The Journal of Accountancy (November 1978):52–59.

Stobaugh, Robert and Daniel Yergin, ed. *Energy Future* (New York, Random House: 1979).

Stobaugh, Robert and Daniel Yergin. "The Energy Outlook: Combining the Options," *Harvard Business Review* (January-February 1980):57–73.

Stock, James R. and Douglas M. Lambert. "The Corporate Energy Plan," *California Management Review* (Fall 1979):5–12.

Swidler, Joseph C. "The Elements of a Sound National Energy Policy," *Public Utilities Fortnightly* (January 3, 1980):13–16.

Uhler, Robert G. and Benjamin Zycher. "Energy Forecasting and Its Uncertainties," *Public Utilities Fortnightly* (January 17, 1980):27–33.

Vernon, Raymond, ed. *The Oil Crisis* (New York, Norton and Company:1976).

Willrich, Mason. "America's Energy Future: Political and Institutional Issues," *Public Utilities Fortnightly* (January 17, 1980):17–20.

National Association of Accountants
Committee on Research
1981–82

Calvin A. Vobroucek
Chairman
Caterpillar Tractor Company
Peoria, Illinois

George Bannon
Moravian College
Bethlehem, Pennsylvania

Robert U. Boehman
Jasper Rubber Products
Jasper, Indiana

Germain Böer
Vanderbilt University
Nashville, Tennessee

Willard Cox
Oil City Iron Works
Corsicana, Texas

Paul E. Dascher
Drexel University
Philadelphia, Pennsylvania

Dwight H. Davis
A. O. Smith Corporation
Kankakee, Illinois

Paul H. Davis
Solar Tubines Incorporated
A Subsidiary of
Caterpillar Tractor Company
Encinitas, California

Geraldine F. Dominiak
Texas Christian University
Fort Worth, Texas

Patricia P. Douglas
University of Montana
Missoula, Montana

J. Bernard Eck
May Zima & Company
Daytona Beach, Florida

Homer R. Figler
CBN University
Virginia Beach, Virginia

Theodore C. Gearhart
U.S. General Accounting Office
Arlington, Virginia

Charles L. Grant
Becton Dickinson & Company
Rochelle Park, New Jersey

Robert E. Hampel
Keller-Crescent Company
Evansville, Indiana

John L. Hanson
Electro Corporation
Sarasota, Florida

John H. Holzapfel
Coopers & Lybrand
Pittsburgh, Pennsylvania

Henry M. Klein
General Instrument
El Paso, Texas

Thomas J. O'Reilly
Coopers & Lybrand
Cleveland, Ohio

Ronald J. Patten
University of Connecticut
Storrs, Connecticut

W. Peter Salzarulo
Miami University
Oxford, Ohio

Henry A. Schwartz
IBM
Armonk, New York

James W. Truitt
Martin Industries, Inc.
Florence, Alabama

Milton F. Usry
Oklahoma State University
Stillwater, Oklahoma

George W. Van Dam
Vanmark Corporation
Creston, Iowa

Robert N. Williams
First National Bank
Dalton, Georgia

Thomas H. Williams
University of Wisconsin
Madison, Wisconsin